ENCYCLOPAEDIC DICTIONARY OF SCIENCE

ENCYCLOPAEDIC DICTIONARY OF SCIENCE

Vol. 1

[A—C]

Edited by
SATISH ANAND

ANMOL PUBLICATIONS PVT. LTD.
New Delhi-110 002

ANMOL PUBLICATIONS PVT. LTD.
4374/4B, Ansari Road, Daryaganj
New Delhi-110 002

Encyclopaedic Dictionary of Science

First Edition 1995
Reprint 1997
ISBN 81-7488-003-8(Set)

PRINTED IN INDIA

Published by J.L. Kumar for Anmol Publications Pvt. Ltd., New Delhi-110 002 and Printed at Mehra Offset Press, Delhi.

Preface

The Encyclopaedic Dictionary of Science is an effort to keep pace with continuing rapid developments and expanding vocabularies in various branches of science. This dictionary has been encyclopaedic in its contents and style of representation. Throughout the process of compilation and editing the volumes, the main effort was to make the dictionary serve as a ready reference for undergraduate and postgraduate students of science and for those appearing for competitive examinations, research scholars, teachers and authors. The entries have been written in a clear and explanatory style to provide both straight forward definitions and invaluable background information. This approach, combined with an extensive cross-reference system, enables the reader to place the each entry into a broader scientific context. At certain appropriate places, the diagrams have been included whenever the meaning of a word can be best understood by means of a diagram.

The presentation throughout has been aimed at sustaining the interest of readers while enriching their vocabulary and comprehension of technical terms and expressions of various branches of science.

This encyclopaedic dictionary will be of immense value to the students of science and to scientists and research scholars working on science projects.

In compiling a set of this kind it becomes necessary to draw upon the work of many authorities and seek the advice of colleagues to all of whom the editor is deeply indebted. All the comments from users on omissions or shortcomings will be most welcome.

The Editor

Vol. 1

A — C

Aardvark. The nocturnal Mammal (genus *Orycteropus*) found in Africa. About 6 ft (180 cm) long, it has a long snout, large, erect ears, a body almost devoid of hair, and a long tail. It claws open ant and termite nests with its forefeet and uses its long, sticky tongue to capture insects.

Ab-. Prefix placed in front of the name of a practical electrical unit to name the corresponding electromagnetic unit. Thus the e.m.u. of current is the abampere.

Abacus. An ancient computing device using movable beads strung on a number of parallel wires within a frame. Each wire represents a decimal place. Ones, tens hundreds, and so on. The beads are grouped to form numbers and shifted in specified patterns to add, subtract, multiply, or divide.

Abalone. Marine Gastropod mollusk (genus *Haliotis),* covered by a single ear-shaped shell perforated with respiratory holes on one side. The abalone is hunted for its large, edible muscular foot and the iridescent Mother-of-Pearl lining of its shell, used for buttons. It feeds by scraping the substrate with its rasping tongue (radula).

Abbe condenser. A simple two-lens system used as a condenser in microscopes. [After Ernst Abbe (1840-1905), German physicist.]

Abdomen. In vertebrates, portion of the trunk between the diaphragm and lower pelvis. In humans the abdominal cavity is lined with a thin membrane, the peritoneum, which encloses the

Stomach, intestines, Liver, and Gall Bladder. The Pancreas, Kidneys, urinary bladder, and, in the female reproductive organs are also located within the abdominal cavity. In insects and some other invertebrates the term *abdomen* refers to the rear portion of the body.

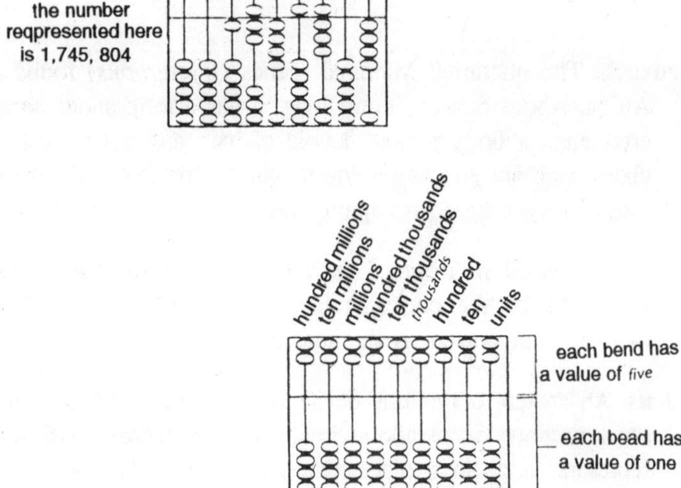

Fig. A-1. Chinese abacus: Numbers are represented by moving beads to the central crossbar.

Aberration. In optics, condition that causes a blurring and loss of clearness in the images produced by lenses or mirrors. Spherical aberration is the failure of a Lens or Mirror of spherical section to bring parallel rays of light to a single focus; it can be prevented by using a more complex parabolic section. Chromatic aberration, the blurred colouring at the edge of an image, arises because some colors of light are bent, or refracted, more than others after passing through a lens; it can be cured by using a corrective lens.

Aberration of Starlight. Angular displacement, caused by the earth's orbital motion, of the apparent path of light from a star, resulting in a displacement of its apparent position from its true position.

Abominable snowman or yeti. Manlike creature associated with the Himalayas. Known only through tracks ascribed to it and alleged encounters, it is supposedly 6 to 7 ft (1.8 to 2.1 m) tall and covered with long hair. While many scholars dismiss it as a myth, others claim that it may be a kind of ape.

Abortion. Expulsion of the embryo or fetus before it is viable outside the uterus, i.e., before the 28th week after conception, in humans (see Reproduction). Spontaneous abortion, or miscarriage, may be caused by death of the fetus due to abnormality or disease or by trauma to the expectant mother. Abortion may also be induced, the fetus removed from the uterus by such procedures as vacuum suction, dilation and curettage, intrauterine saline injection, and hysterotomy (surgical incision of the uterus). Abortion was long practiced as a form of Birth Control until pressure from the Roman Catholic Church and changing opinion led in the 19th cent. to the passage of strict antiabortion laws (e.g. in England and the U.S.). Attitudes toward abortion have generally become more liberal in the 20th cent. By the 1970s, abortion had been legalized in most European countries, the USSR, and Japan; in the U.S., according to a 1973 Supreme Court ruling (see Roe V. Wade), abortions are permitted during the first six months of pregnancy. Abortion remains a controversial issue in the U.S. however, and in 1977 Congress barred the use of Medicaid funds for abortion except for therapeutic reasons and in certain other specified instances.

Abrasive. Material used to grind, smooth, cut, or polish another substance. Natural abrasives include Sand, Pumice, Corundum, and ground Quartz. Carborundum (Silicon Carbide) and Allumina (aluminum oxide) are major synthetic abrasives. The hardest abrasives are natural or synthetic diamonds, used in the form of dust or minuscule stones.

Abscess. Accumulation of pus in tissues as a result of infection. Characterized by inflammation and painful swelling, it may occur in various parts of the body, e.g., skin, gum, eyelid (sty), and middle ear (mastoid infection). Many abscesses respond to treatment with Antibiotics; others require surgical drainage.

Abscissa. The x coordinate of a point on a two-dimensional Cartesian graph: i.e. the distance of a point from the y axis, measured along the x axis from the origin; a value of the independent variable of a function of x. *Compare* ordinate.

Absolute Alcohol. Pure alcohol: i.e. ethanol containing little or no water.

Absolute value. Magnitude of a mathematical expression, disregarding its sign; thus the absolute value is always positive. In symbols, if $|a|$ denotes the absolute value of a number a, then $|a| = a$ for $a > 0$ and $|a| = -a$ for $a < 0$.

Absolute zero. The zero value of thermodynamic temperature, equal to 0 kelvin or - 273.15 C. It may be regarded as the temperature at which molecular motion ceases.

Absorptance (absorption Factor). Symbol: α. A measure of the extent to which something can absorb radiation, equal to the ratio of the absorbed flux to the flux incident on the body. For a black body, the absorptance is 1. The quantity was formerly called the *absorptivity.*

Absorption. **1.** Solution of a gas in a solid or liquid. The gas permeates into the bulk of the material: this process is distinguished from *absorption,* in which the gas in held on the surface. Absorption in solids is sometimes called *sorption.*

2. Conversion of the energy of electromagnetic or sound waves into some other form of energy in a medium. A beam of light, for example, passing through matter loses intensity. Part of the loss may result from excitation of atoms or molecules.

This process is distinguished from *diffusion,* in which light is scattered out of the beam. In absorption, the process occurring depends on the way the radiation interacts with the material: infrared radiation, for instance, is covered directly into heat by exciting vibrations of the atoms.

Absorption. Taking a molecules of one substance directly into another substance. Absorption may be either a physical or a chemical process. Physical absorption depends on the solubility of the substance absorbed, and chemical absorption involves chemical reactions between the absorbed substance and the absorbing medium.

Absorption spectrum. A spectrum produced by absorption of electromagnetic radiation by matter. In producing an absorption spectrum, a continuous source of radiation is used: i.e. one having a range of wavelengths. This is passed through the sample and the emerging beam is dispersed using a prism or grating. The electromagnetic radiation is absorbed at certain frequencies, these being capable of exciting the atoms or molecules of the sample from their ground state to an excited state. Thus, in the case of visible light, the absorption spectrum consists of dark lines or bands on the bright continuous background. The frequency (ν) at which a line occurs depends on the difference ΔE between the energy of the ground and that of the excited state: $\Delta E = h\nu$, where h is the Planck constant. See also emission spectrum, Fraunhofer lines.

Absorptivity. *See* absorptance.

Abundance. 1. Symbol: *C*. The concentration of given isotope in a mixture of isotopes, equal to the ratio of the number of atoms of the isotope to the total number of atoms; often expressed as a percentage.

2. The concentration of a specified element in the earth's crust, the universe, etc.

A.C. *See* alternating current.

Acacia. Plant (genus *Acacia)* of the Pulse family, mostly tropical and subtropical thorny shrubs and trees. Some have a feathery foliage composed of leaflets; others have no leaves but have flattened leaflike stems containing chlorophyll. Various species yield lac (for shellac), catechu (a dye), gum arabic, essential oils, tannins, and hardwood timber.

Acanthus. Common name for the Acanthaceae, a family of chiefly perennial herbs and shrubs, mostly tropical. Many members have decorative spiny leaves and are cultivated as ornamentals, e.g. bear's breech, whose ornate leaves provided a motif often used in Greek and Roman art and architecture. In Christian art, the acanthus symbolizes heaven.

Acceleration. The rate of change of velocity with time. Linear acceleration (symbol: *a*) is measured in metres per second per second, etc. Angular acceleration (symbol : ω) is measured in radians per second per second, etc.

Acceleration of free fall. Symbol: g. The acceleration of a body falling freely, i.e. with no air resistance, at a specified point on the earth's surface as a result of the gravitational attraction of the earth. This acceleration, which is often called the *acceleration due to gravity,* varies from place to place because of different distances from the earth's surface to its centre. The standard value of *g* is 9.806 65 m s^{-2} (32.174 ft 5^{-2}).

Accelerator. 1. A machine for increasing the velocity (and thus the energy of charged elementary particles and ions. Accelerators are used in fundamental research in nuclear and particle physics for producing a high-energy beam of particles which is directed onto a target, usually within a bubble chamber. The particles are initially produced from a hot filament (in the case of electrons) or from an ion source.

The simplest method of accelerating charged particles is to use a high potential difference, the energy being gained by the particle in the electric field. This method is used in the

Van de Graaff accelerator. Higher energies are achieved by increasing the particle velocity in stages by a series of small electric fields—the method used in linear accelerators, in which the particles follow a straight path down an evacuated tube. A similar technique is used in the cyclotron and synchro-cyclotron, which employ magnetic fields to cause the particles to travel in spiral paths.

Cyclic accelerators of this type have the advantage that long path lengths can be used without making the accelerator impractically large. The betatron is a cyclic accelerator in which energy is gained by magnetic induction produced by a varying magnetic field. Synchrotrons work by a combination of electric and varying magnetic fields. The proton synchrotron is the most powerful type of accelerator, producing protons in the GeV range.

Particle energies can be effectively increased by the use of *storage rings,* which are large evacuated toroidal rings into which the particles are injected. It is possible to build up an intense beam of particles and maintain it circulating around the ring for many months. It is also possible to produce two beams, as of electrons and positrons, travelling in opposite directions and to arrange for them to intersect in the ring. In this way particle interactions can be studied at energies of up to 1700 GeV.

2. A substance added to increase the rate of a chemical reaction i.e. to act as a catalyst. Accelerators are used to increase the rate of vulcanization of rubber.

Acceptor. 1. An atom that accepts electrons in an extrinsic semiconductor, thus producing holes in the conduction band.

2. The atom or molecule that contributes no electrons in forming a coordinate bond. *Compare* donor.

Access Time. The time necessary for information to be supplied from a computer store for processing.

Accommodation. *See* eye.

Accumulator (storage battery). A device designed for storing electricity, consisting of one or more secondary cells in series. Common examples are the lead-acid accumulator and the nife cell.

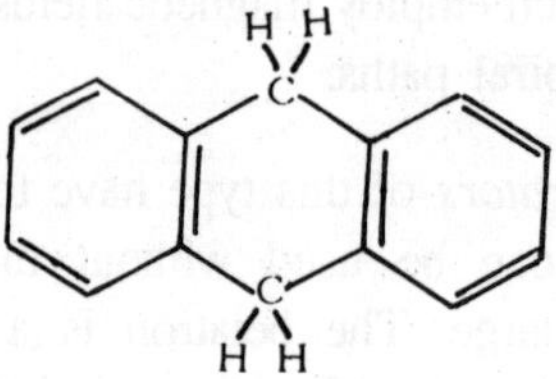

Fig. A-2. Acenaphthene.

Acenaphthene. A white crystalline solid, $C_{10}H_6$ $(CH_2)_2$, obtained from coal tar and used in the manufacture of dyes and plastics. M.pt. 96.2° C; b, pt. 279° C; r.d. 1.02.

Acetal. Any of a class of organic compounds with the general formula RCH (OR') OR''), where R, R', and R'' are organic groups. Acetals are formed by addition of an alcohol to an aldehyde. The reaction proceeds in two steps. The first is addition of one alcohol molecule to form a *hemiacetal,* with further reaction forming the full acetal. The reaction is catalysed by acid.

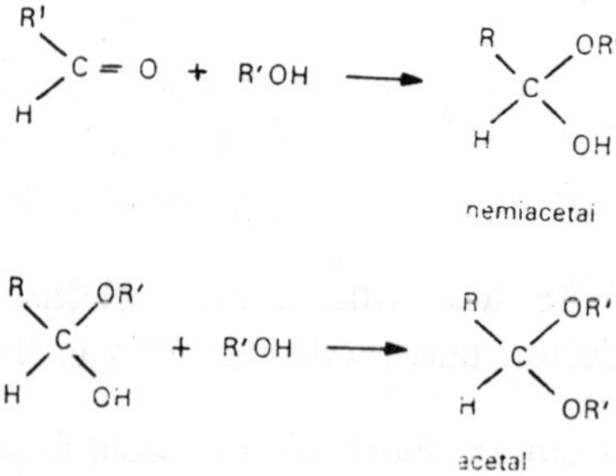

hemiacetal

acetal

Fig. A-3. Acetal formation.

Ketones react similarly to yield ketals. The term *acetal* is often applied to *1,1 - diethoxyethane,* $CH_3CH(OC_2H_5)_2$, B.pt. 103.2° C; r.d. 0.83.

Acetaldehyde (ethanal). A colourless liquid aldehyde, CH_3CHO, made by the oxidation of ethanol and used in manufacturing many other compounds. M.pt. -124.6 C; b.pt. 20.8° C; r.d. 0.78.

Acetaldol (aldol, 3-hydroxybutanal). A courless syrupy liquid, $CH_3C\text{-}HOHCH_3CHO$, made by the condensation of acetaldehyde and used in medicine as a sedative and hypnotic. M.pt. below 0° C; b.pt. 83° C; r.d. 1.11.

Acetamide (ethanamide). A colourless deliquescent crystalline amide, $CH_3C\text{-}ONH_2$, with a mouse-like odour. It is manufactured from ethyl acetate and ammonia and used as a solvent and wetting agent. M.pt. 82.3° C; b.pt, 22.2° C; r.d. 1.0.

Acetanilide. An odourless white powder, $C_6H_5NH\ (COCH_3)$, made by the acetylation of aniline and used in the preparation of drugs, dyes, and lacquers. M. pt. 114.3°C; b.pt. 304° C; r.d. 1.22.

Acetate. 1. Any salt or ester of acetic acid. 2. See cellulose acetate.

Acetic acid (ethanoic acid). A colourless liquid carboxylic acid, CH_3COOH, manufactured by bacterial oxidation of ethanol (see vinegar) or by the oxidation of acetaldehyde. The anhydrous liquid is known as *glacial* acetic acid. Acetic acid is used in the manufacture of dyes, rebbers, plastics, and many other products. M.pt. 16.6° C; b. pt. 117.9° C; r.d. 1.05.

Acetic anhydride (ethanoic anhydride). A colourless liquid organic anhydride, $CH_3CO)_2O$, used as a dehydrating and acetylating acetylating agent (*see* acetylation) M.pt. -73.1° C; b.pt. 139.6° C; r.d. 1.08.

Acetolysis. The reaction of an organic compound with glacial acetic acid to exchange one of its groups for an acetyl group. A tertiary alkyl halide, for example, reacts to give an ester; $(CH_3)_3CCl + CH_3COOH = CH_3COOC(CH_3)_3 + HCL$. Acetolysis is a types of solvolysis.

Acetone, dimethyl ketone. or **2-propanone** CH_3COCH_3), colorless, flammable liquid. Acetone is widely used in industry as a solvent for many organic substances and is a component of most paint and varnish removers. It is used in making synthetic resins and fillers, smokeless powders, and many other organic compounds.

Acetonitrile (methyl cyanide). A colourless flammable liquid nitrile, CH_3CN, manufactured by the dehydration of acetamide and used in the synthesis of some pharmaceuticals. M.pt. -45.7° C; b.tp, 81.6° C; r.d. 0.79.

Acetoplenone (phenyl methyl ketone). A colourless liquid ketone, $C_6H_5COCH_3$, with a sweet pungent odour and taste, used in perfumery. M.pt. 20.5° C; b.pt. 202° C; r.d. 1.03.

Acetylation. The introduction of an acetyl group CH_2CO-) into an organic compound in a chemical reaction. Acetylations are thus a special type of acylation reaction. Most acetylations involve replacement of the hydrogen atom of a hydroxyl group (-OH) or amino group ($-NH_2$) by the acetyl group, Common acetylating agents are acetic anhydride, acetyl chloride, and ketene.

Acetyl chloride (ethanoyl chloride). A colourless flammable pungent liquid acid chloride, CH_3COCl, used as an acetylating agent (*see* acetylation). M.pt -112°C: b. pt. 50.9° C; r.d. 1.11.

Acetylcholine. Organic compound containing carbon, hydrogen, oxygen, and nitrogen, essential for the conduction of nerve impulses in animals. It is found in highest concentrations on neuron surfaces and is liberated at nerve cell endings. There is strong evidence that acetylcholine is the transmitter substance that conducts impulses from one cell to another in the parasympathetic nervous system, and from nerve cells to smooth muscle, skeletal muscle, and exocrine glands.

Acetylene or ethyne. ($HC \equiv CH$).a colorless gas and the simplest alkyne (see Hydrocarbon). Explosive on contact with air, it is stored dissolved under pressure in Acetone. It is used to make neoprene Rubber, Plastics, and Resins. The oxyacetylene torch mixes and burns oxygen and acetylene to produce a very hot flame — as high as 6300°F (3480°C)— that can cut steel and weld iron and other metals.

Acetyl group. The organic group CH_3CO-.

Acetylide. A carbide that yields acetylene on hydrolysis.

Acetylsalicylic acid. *See* asprin.

Achromat. *See* achromatic.

Achromatic. 1. Having no colour *Achromatic colours* are colours such as black, white, and various greys, which have no hue.

2. Having no chromatic aberration. An *achromatic lens* is one made by combining lenses of different types of glass, so that the combination has no dispersion. The simplest type, an *achromatic doublet,* is made of two lenses of flint and crown glass. The power of the doublet is the sum of the powers of the lenses ($P = P_1 + P_2$). For the combination to be achromatic, the lenses must satisfy the relationship $P_1\omega_1 + P_2\omega_2 = 0$, where ω_1 and ω_2 are the dispersive powers of the glasses used. Achromatic lenses are called *achromats.*

Acid. A substance that will react with a base to form a salt and water. Acids turn litmus red and produce hydrogen ions if dissolved in water. They react with certain metals to give hydrogen and a metal salt, *Strong acids* are substances such as hydrochloric acid, in which the covalent molecule is fully dissociated into ions in solution: $HCl = H^+ Cl^-$, *Weak acids* are compounds such as acetic acid, which are partially dissociated. In solution the hydrogen ion, H^+, is solvated by water and often considered to be a hydroxonium ion, H_3O^+. The idea of an acid in chemistry has been extended to include Lewis acids.

Acid chloride (acyl chloride). Any of a class of organic compounds with the general formula RCOCl, in which R is a hydrocarbon group. They are derived by replacing the hydroxyl group of a carboxylic acid by a chlorine atom using a chlorinating agent such as phosphorus trichloride. Acid chlorides, such as acetyl chloride, $CH_3COCl + H_2O = CH_3COOH + HCl$. They similarly react with alcohols and other hydroxy compounds and are used as acylating agents (see acylation). Other *acid halides* (or *acyl halides)* such as RCOF, and RCOBr, and RCOI, show similar behaviour.

Acid Dye. Any of a class of dyes that consist of salts of alkali metals containing large coloured negative ions. Acid dyes can be used on wool and silk. Azo dyes are examples of acid dyes.

Acid Halide. *See* acid chloride.

Acidic. Denoting a substance that is an acid or a solution containing an excess of hydrogen ions.

Acidic hydrogen. A hydrogen atom in an acid that can be lost to form a hydrogen ion. For example, of the four hydrogen atoms in the acetic acid molecule, CH_3COOH, only the atom joined to the oxygen atom is an acid hydrogen: $CH_3COOH = CH_3COO^- + H^+$.

Acidimetry. Determination of the amounts of acids in solution by titration.

Acidolysis. Hydrolysis of carboxylic esters in acid solution to yield the parent carboxylic acid and the alcohol. The acid acts as a catalyst by protonating the carbonyl oxygen, thus facilitating attack by water.

Acid radical. The group bound to the acidic hydrogen atom or atoms in an acid. For example, the acetate group, CH_3COO^-, is the acid radical of acetic acid.

Acid rain. Form of precipitation (rain, snow, sleet, or hail) containing high levels of sulfuric or nitric acids (pH below 5.5-5.6). Produced when sulfur dioxide and various nitrogen oxides combine with atmospheric moisture, acid rain can contaminate drinking water, damage vegetation and aquatic life, and erode buildings and monuments. It has been an increasingly serious problem since the 1950s, particularly in the NE U.S., Canada, and W Europe, especially Scandinavia. Automobile exhausts and the burning of high-sulfur industrial fuels are thought to be the main causes, but natural sources, e.g., volcanic gases and forest fires, may also be significant. See also Ecology; Pollution, Waste Disposal.

Acids and bases. Two related classes of chemicals; the members of each class have a number of common properties when dissolved in a solvent, usually water. Acids in water solutions exhibit the following common properties : they taste sour; turn Litmus paper red; and react with certain metals, such as zinc, to yield hydrogen gas. Bases in water solutions exhibit these common properties: they tste bitter; turn litmus paper blue; and feel slippery. When a water solution of acid is mixed with a water solution of base, a salt and water are formed; this process, called neutralization, is complete only if the resulting solution has neither acidic nor basic properties. When an acid or base dissolves in water, a certain percentage of the acid or base particles will break up, or dissociate, into oppositely charged ions. The Arrhenius theory of acids and bases defines and acid as a compound that can dissociate in water to yield hydrogen ions (H^+) and a base as a compound that can dissociate in water to yield hydroxyl ions (OH^-).

The Bronsted-Lowry theory defines an acid as a proton donor and a base as a proton acceptor. The Lewis theory defines an acid as a compound that can accept a pair of electrons and a base as a compound that can donate a pair of electrons. Each of the three theories has its own advantages and disadvantages each is useful under certain conditions. Strong acids, such as Hydrochloric Acid, and strong bases, such as potassium hydroxide, have a great tendency to dissociate in water and are completely ionized in solution. Weak acids, such as acetic acid, and weak bases, such as *Ammonia,* are reluctant to dissociate in water and are only partially ionized in solution. Strong acids and strong bases make very good electrolytes (see electrolysis), i.e., there solutions readily conduct electricity. Weak acids and weak bases make poor electrolytes.

Acne. Inflammatory disease of the sebaceous glands, characterized by blackheads, cysts, and pimples. The lesions appear on the face, neck, chest, back, and arms, and may be mild to severe. Most prevalent during adolescence, acne may appear in adulthood. Its cause is unknown, but contributing factors include genetic predisposition and hormonal changes during puberty. Treatment includes use of cleansers, Antibiotics, surgical drainage of lesions, and, for severe cases, retinoic acid derivatives.

Acoustics. The science of Sound, including its production, propagation, and effects. An important practical application of acoustics is in the designing of auditoriums, which requires a knowledge of the characteristics of sound Waves. Reflection of sound can cause an Echo, and repeated reflections in an enclosed space can cause reverberation, the persistence of sound. Some reverberation in auditoriums is desirable to avoid deadening the sound of music. Reflection can be reduced through the proper configuration and texture of walls, and by the use of sound-absorbent materials. Another acoustical problem is Interference, which can create "dead spots" in auditoriums for certain frequencies.

Acquired characteristics, modifications produced in an individual plant or animal as a result of mutilation, disease, use and disuse, or any distinctly environmental influence. Belief in the inheritability of acquired characteristics was accepted by Lamarck but ultimately rejected by modern geneticists, who have affirmed that inheritance is determined solely by reproductive cells and unaffected by somatic (body) cells.

Acre. An Imperial unit of area equal to 4840 sq. yards. It is equivalent to 4046.856 422 sq. metres.

Acriflavine. A brown-orange granular solid, $C_{14}H_{14}N_3Cl$, used as an antiseptic and bacteriostat.

Acrolein (propenal). A colourless or yellowish flammable liquid aldehyde, CH_2 : CHCHO, manufactured from propene and used for producing polyurethane and polyester resins. M. pt. -87°C; b.pt. 53° C; r.d. 0.84.

Acrylic Resin. Any of a class of synthetic resins obtained by polymerization of acrylic acid, methacrylic acid, or their esters. The acrylics are thermoplastic materials noted for their transparency. The commonest is polymethylmeth-acrylate.

Acrylonitrile (vinyl cyanide). A colourless liquid nitrile, H_2C:CHCN, manufactured from propene, air, and ammonia. It is copolymerized with butadiene to make synthetic rubbers and is also used in the production of acrylic fibres. M. pt. -83.5°C; r.d. 0.81.

Actinic. Denoting radiation that causes chemical reaction, especially ultraviolet radiation.

Actinide. Any of 15 elements from actinium (atomic number 89) to lawrencium (atomic number 103) inclusive. The actinides are similar to the lanthanides in that they form a series in which the 5f shell is being filled. Chemically, the actinides are reactive metals forming compounds with a variety of valencies. All are radioactive and those with atomic numbers greater than 92—the *transuranic elements*— are artificial elements, made by bombardment of other nuclei with high-energy particles.

Actinide series, the radioactive metals, with atomic numbers 89 through 103, in group IIIb of the Periodic Table. They are Actinium, Thorium, Protactinium, Uranium, Neptunium, Plutonium, Americium, Curium, Berkelium, Californium, Ensteinium, Fermium, Mendelevium, Nobelium, and Lawrencium. All members of the series have chemical properties similar to actinium. Those elements with atomic numbers greater than 92 are called Transuranium Elements.

Actinium. Symbol: Ac. An actinide element found in uranium ores. Its most stable isotope, ^{227}Ac, has a half-life of 21.7 years. A.N. 89; m.pt. 1050° C; r.d. 10.07; valency 3.

Actinometer. Any of various instruments for measuring the intensity of radiation.

Action. The isotope of radon with mass number 219. Its half-life is 3.92s.

Action. The product of momentum and distance.

Activate. 1. To heat a solid porous material in order to drive off absorbed water and other substances, thus producing an absorbent material. *Activated alumina,* made by heating aluminium oxide, is used for absorbing gases and for chromatography. *Activated charcoal* is also a highly effective absorbent. It is used for removing gas from vacuum systems, separating gases by selective absorption, purifying air in gas masks, decolorizing, deodorizing, and similar applications.

2. To supply sufficient energy to an atom or molecule to make it reactive.

3. To make a material radioactive, as by bombardment with neutrons or other particles.

Activated complex. *See* transition state.

Activation analysis. A technique for determining the amount of a particular element in a sample by bombarding the sample to produce a radioactive isotope, which is then identified by its gammaray emission spectrum. A common method of activating the sample is by placing it in a nuclear reactor, where radioisotopes are formed by neutron capture. The technique is extremely sensitive.

Activation energy. The energy required to initiate a reaction. In a chemical reaction the activation energy is the energy that must be supplied in breaking and reforming chemical bonds; i.e. it is the energy required to form the transition state. *See also* Arrhenius equation.

Active mass. The effective concentration of a substance in a chemical reaction, as used in the law of mass action.

Activity. Symbol : A. A measure of the radioactivity of a radioactive substance, equal to the number of atoms disintegrating per unit time. It is measured in curies.

Acupuncture. Technique of medical treatment, based on traditional Chinese medicine, in which a number of very fine metal needles are inserted into the skin at any of some 800 specially designated points. In China it has long been used for pain relief and treatment of such ailments as Arthritis, Hypertension, and Ulcers, More recently it has been used as an Anesthesia for childbirth and some surgery (Unlike conventional anesthesia, it does not lower blood pressure or depress breathing). It has been suggested that acupuncture works, by stimulating the production of endorphins. U.S. research on acupuncture has focused on its use in pain relief and anesthesia.

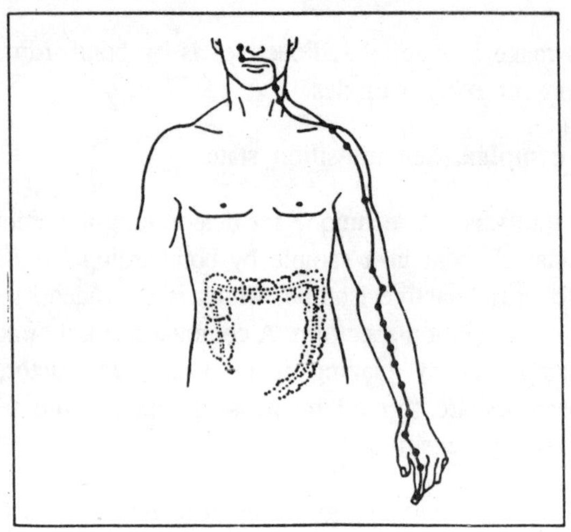

Fig. A-4. Acupuncture points for treating diseases of the large intestine.

Acute. Denoting an angle that is between 0° and 90°.

Acyclic. Denoting a chemical compound that is not cyclic, i.e. does not contain a ring of atoms in its molecular structure. Propane and acetic acid are common examples of acyclic compounds. *Compare* cyclic.

Acylation. The introduction of an acyl group (RCO-) into an organic compound in a chemical reaction. The most important acylations are those in which an acyl group is attached to an aromatic ring by a Friedel-Crafts reaction. *See also* acetylation.

Acyl chloride. See acid chloride.

Acyl group. Any organic group with the formula RCO-, where R is a hydrocarbon group. Examples are the acetyl group, CH_3CO- and the benzol group, C_6H_5CO-.

Acyl halide. *See* acid chloride.

Adaptation. In biology, the adjustment of living matter to environmental conditions, including other living things. Animals and plants are adapted for securing food and surviving even in conditions of drought, great heat, or extreme cold. Adaptations are believed to arise when genetic variations that increase an organism's chances of survival are passed on to succeeding generations.

Adaptive Radiation. In biology, the evolution of an ancestral species adapted to a particular way of life into several species, each adapted to a different habitat. Illustrating the principle are the 14 species of Darwin's finches, small land birds of the Galapagos Island: 3 are ground-dwelling seedeaters, 3 live one cactus plants and are seedeaters, 1 is a tree-dwelling seedeater, 7 are tree-dwelling insect eaters, but all derive from a single species of ground-dwelling, seedeating finch that probably emigrated from the South American mainland.

Addend. One of a set of numbers to be added.

Addison's disease. Progressive disease brought about by atrophy of the outer layer (cortex) of the adrenal gland; also called chronic adrenocortical insufficiency. The deterioration of this tissue causes a decrease in the secretion of vital steroid hormones, producing such symptoms as Anemia, weakness, abnormal skin pigmentation, and weight loss. The cause of the disease is unknown. Once thought inevitably fatal, the disease is treated with adrenocortical hormones that enable its victims to lead a nearly normal life.

Addition. A chemical reaction in which one molecule or atom combines with another to form a third molecule. In organic chemistry, addition is usually a reaction in which a molecule adds to an unsaturated molecule by breaking a double or triple bond. Alkenes, alkynes, and aldehydes all undergo addition reactions.

Addition polymerization. *See* polymerization.

Additive process. The process in which a coloured light is produced by direct combination of lights of different colours. Most colours can be produced by a suitable mixture of primary colours. *Compare* substrative process.

Adenosine Triphosphate. (ATP) Compound composed of adenine (a purine), ribose, and three phosphate units; one of the most important low-molecular-weight molecules in living matter. It is a rich source of the chemical energy necessary for the success of a vast number of chemical reactions in the cell. ATP also plays a role in kidney function, transmission of nerve impulses, muscle contraction, active transport of materials through cell membranes, and synthesis of Nucleic Acids and other large molecules.

Adhesion and Cohesion. Attractive forces between material bodies. Adhesive forces act between different substances, whereas cohesive forces act within a single substance, holding its atoms, ions, or molecules together. Were it not for these forces, solids and liquids would act as gases. Surface Tension in liquids results from cohesion, and Capillarity results from a combination of adhesion and cohesion. Friction between two solid bodies depends in part on adhesion.

Adiabatic. Denoting a process in which no heat enters or leaves the system. In general, the temperature changes when an adiabatic change occurs; for example, in the adiabatic compression of a gas work is done on the system and the gas temperature rises. *Compare* isothermal.

Adiabatic demagnetization. A technique for producing very low temperatures (close to absolute zero) by demagnetizing a paramagnetic salt. The magnetized material is first cooled to the temperature of liquid helium and isolated thermally form its surroundings. When the field is removed the sample is demagnetized and cools still further.

Adiabatic equation. The equation $p^{V\gamma} = C$, describing the relation between the pressure and volume of a gas during an adiabatic change. C is a constant and Y is the ratio of the principal specific heat capacities of the gas.

Adipic acid (hexanedioic acid). A colourless crystalline carboxylic acid , $HOOC(CH_2)_4COOH$, manufactured by oxidation of cyclohexanol. It is used in the manufacture of synthetic polymers, particularly nylon. M.pt. 153° C; b.pt. 267° C (750 mmHg); r.d. 1.4.

Admittance. Symbol: Y. A measure of the ability of a circuit to pass electric current, equal to the reciprocal of its impedance. It is a complex quantity: $Y = G + iB$, where G is the conductance and B is the susceptance. Admittance is measured in siemens.

Adolescence. Time of life from onset of puberty to full adulthood. Falling approximately between the ages of 12 and 21, adolescence is characterized by physical changes leading to sexual maturity, problems of sex-role identification and achievement, and movement toward personal independence. Psychologists regard adolescence as a period of social pressure specifically related to the society, not as a unique biological period.

Adrenal glands. Pair of small endocrine glands (see Endocrine system) situated atop the kidneys. The outer yellowish layer (cortex) secretes about 30 steroid hormones, most importantly aldosterone, which regulates water and salt balance in the body, and cortisol, which controls carbohydrate, fat, and protein metabolism. The inner reddish portion (medulla) of the adrenals secretes the emergency-response hormones Epinephrine (adrenaline) and norepinephrine.

Adsorption. The process in which a compound, usually a gas, forms a layer on the surface of a solid: this process is distinguished from *absorption.* in which the gas permeates into the material. Two types of adsorption are distinguished: *chemisorption,* in which the gas molecules or atoms are held by covalent bonds, and *physisorption,* in which they are held by the weaker van der Waals forces. In chemisorption, a single layer of adsorbed molecules is formed : physisorption may involve the production of several layers.

Aerial (Antenna). A device for transmitting and receiving radio waves. A simple form is the *dipole aerial*— a long metal rod divided at the centre point with connections made at this point. The overall length of the dipole is about one half of the operating wavelength. In transmitting, the electromagnetic waves are produced by accelerating charges in the aerial. In receiving, the electromagnetic radiation induces small varying currents. Many different types of aerial exist suitable for use at various frequencies. Arrays of elements are used for reception and transmission in particular directions.

Aerodynamics. Study of gases in motion. Because the principal application of aerodynamics is the design of Airplanes, air is the principal gas with which this science is concerned Bernoull's principle, which states that the pressure of a moving gas decreases as its velocity increases, has been used to explain the lift produced by a wing having a curved upper surface and a flat lower surface (*see* Airfoil). Because the flow is faster across the curved surface than across the plane one a greater pressure is exerted in the upward direction. Aerodynamics is also concerned with the drag caused by air friction, which is reduced by making the surface area of the craft as small as possible. At speeds close to the speed of sound, or Mach 1 (see Mach number), there is also a large, sudden increase of drag, which has been called the sonic, or sound, barrier. Aerodynamics is also used in designing automobile bodies and trains for minimum drag and in computing wind stresses on bridges, bindings, and the like. The Wind Tunnel is one of the basic experimental tools of the aerodynamicist. See Shock Wave, Sonic Boom.

Aerosol. A Colloid in which small solid or liquid particles are suspended in a gas. Natural aerosols such as fog or smoke occur throughout the Atmosphere, which is itself an aerosol. The term is also used to describe a container of paint, insecticide, or other substance held under pressure by a propellant, which releases the substance in the form of a fine spray or oam. The fluorocarbon Freon was until recently the most common aerosol propellant, but its use has diminished because it is thought to contribute to the destruction of the Ozone layer of the stratosphere.

African Violet. Common name for plants (mostly hybrids of *Saintpulia ionatha*) of the Generia family grown chiefly as houseplants for their colorful flowers and fuzzy foliage.

Afterdamp. Carbon dioxide produced in coal mines by an explosion of firedamp.

Agar. Product consisting of the Sugar galactose, obtained from some red Algae species (see seaweed). Dissolved in boiling water and cooled, agar becomes gelatinous; it is used as a culture medium (especially for Bacteria), a laxative, and a food thickener.

Agate. Extremely fine-grained variety of Chalcedony, banded in two or more colours. The banding occurs because agates are formed by the slow deposition of silicia from solution into cavities of older rocks. Agates are found primarily in Brazil, Uruguay, India, Mexico, and the U.S. They are valued as semiprecious Gems and are used in the manufacture of grinding equipment.

Aggregation. A cluster or group of particles held together in a gas or liquid by intermolecular forces.

Agonic line. A line on the earth's surface joining points of zero magnetic declination: i.e. points at which a compass indicates true north. Two main agonic lines exist; one passing from north to south through America and the other having an irregular path through eastern Europe, Arabia, Asia, and Australia.

Agriculture. Science of producing crops and livestock; it aims to increase production and protect the land from deterioration. Branches include Agronomy, Horticulture, Entomology, animal husbandry, and Dairying. Historically, agriculture has been linked with social, economic, and political organization. Its development among early peoples encouraged stable settlements, and it later became associated with landholding, Slavery, and Feudalism. During the so-called agricultural revolution (16th and 17th cent.), horticultural knowledge greatly expanded, and crops and farming methods were exchanged internationally. The invention of machines such as McCormick's reaper (1831) and the the general-purpose tractor (c.1924) led to mechanized, large-scale farming. Modern Western agriculture also depends on breeding programs; Fertilizers and Pesticides; food processing techniques, e.g., Refrigeration; efficient marketing; agricultural colleges and research centers; and , in many nations, government subsidies.

Agronomy. Branch of Agriculture concerned with soil management and with the breeding, physiology, and production of major field crops. It deals mainly with large-scale crops (e.g., cotton, soybeans, and wheat), while Horticulture concerns fruits, vegetables, and ornamentals.

Ailanthus. Tree (genus *Ailanthus*) of the family Simarubaceae. Ailanthus is native to warm regions of Asia and Australia. Its wood is used in cabinetmaking and for charocal manufacture. The bark and leaves are used medicinally, and the leaves provide food for silkworms. Females of a species called tree of heaven are cultivated for their attractive foliage and their resistance to smoke and soot; the flowers of the male plant, however, have a disagreeable odor.

Air Conditioning. Mechanical process for controlling the temperature, humidity, cleanliness, and circulation of air in buildings and rooms. Most air conditioners operate by ducting air across the colder, heat-absorbing side of a Refrigeration apparatus, and directing the cooled air back into the air-conditioned space. Small window conditioners vent heat outdoors. Larger systems use circulating water or remove heat. Air conditioning provides the heat, humidity, and contamination controls essential in the manufacture of such products as chemicals and pharmaceuticals.

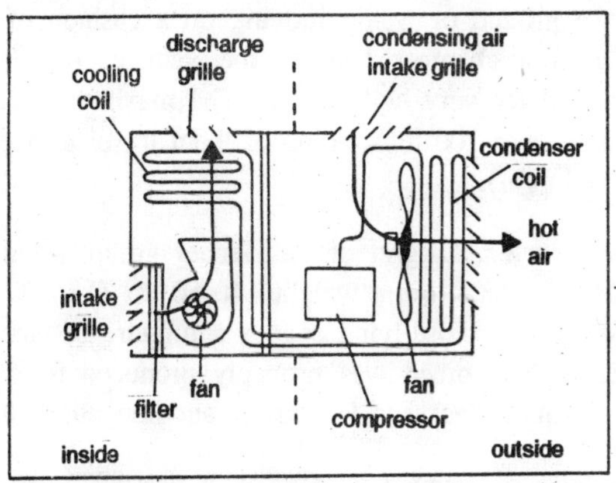

Fig. A-5. Cross section of air conditioning unit.

Aircraft Carrier. Ship designed to carry aircraft and to permit takeoff and landing of planes. Its distinctive features are a flat upper deck (flight deck) that functions as a takeoff and landing field, and a main deck (hangar deck) beneath the flight deck for storing and servicing the aircraft. The aircraft carrier remained an experimental and untested war vessel until World War II, when the Japanese wreaked havoc on the British, Dutch, and U.S. navies with carrier-borne aircraft. By 1942 the aircraft carrier had replaced the battleship as the major unit in a modern fleet, and during the war it was indispensable in naval operations against a sea or shore-based enemy, with two major battles (Coral Sea and Midway, 1942) being fought entirely by aircraft, and the opposing fleets never coming within gunshot range of each other. U.S. carriers of the *Essex* class spearheaded the island-hopping campaign in the Pacific. A new era in carrier design opened when the U.S. launched (1960) the nuclear-powered *enterprise,* a vessel capable of lengthy voyages without refueling. A nuclear-powered successor, the *Nimitz* (1975), was the largest ship afloat when launched in 1975.

Air-cushion vehicle (ACV).**Ground-effect machine,** or **Hovercraft.** Vehicle designed to travel at a short distance above ground or water, moving on a cushion of air that is held in a chamber beneath the vehicle. ACV's offer the potential for very high speeds. The maximum size of ACVs is now over 100 tons; some of them travel at over 100 mph (160 km/hr).

Airedale Terrier. Largest of the Terrier group; shoulder height, c.23 in. (58.4 cm); weight, 40-50 lb (18.1-22.7 kg). Its dense, wiry, close-lying coat is a mixture of tan, black, and grizzle. The breed was probably produced from crosses of the extinct black-and-tan terrier and the ottorhound.

Airfoil. Surface designed to develop a desired force by reaction with a fluid, especially air, that is flowing across the surface. Examples of airfoils are the fixed wings of Airplanes, which produce lift (*see* Aerodynamics), and control surfaces, such as ailerons, elevators, rudders, and flaps, that are manipulated to produce variable forces. Other airfoils include spoilers, propeller blades, and the blades utilized in turbojet engines.

Airplane, Aeroplane, or **Aircraft.** Heavier-than-air vehicle, mechanically drived and fitted with fixed wings that support it in flight through the dynamic action of the air. On Dec. 17, 1903, American Orville and Wilbur Wright flew the first airplane near Kitty Hawk, N.C. the machine was a biplane with two propellers chain-driven by a gasoline motor. Modern airplane arc monoplanes (airplanes with one set of wings). Airplanes may further be classified as driven by propeller, Jet Propulsion, Or Rocket. The airplane has six main parts: fuselage, wings, stabilizer (or tail plane), rubber, one or more engines, and landing gear. The fuselage is the main body, usually streamlined in form. The wings are the main supporting surfaces. The airplane's lift, or force supporting it in flight, is basically the result of the direct action of air against the surfaces of the wings. With the use of jet engines and the resulting higher speeds, airplanes have become less dependent on large values of lift from the wings. Consequently wings have been shortened and swept back so as to produce less drag, especially at supersonic speeds. At the trailing edge of the wings are attached movable surfaces, called airlerons that are used to gain lateral control and to turn the plane. Directional stability is provided by the tail fin, a fixed vertical airfoil at the rear of the airplane. The stabilizer is a fixed horizontal airfoil at the rear of the airplane used to suppress undesired pitching motion. The elevators, which are movable auxiliary surfaces attached to the stabilizers, are used to produce controlled pitching. The rudder, generally at the rear of the tail fin, is a movable auxiliary airfoil that gives the craft a yawing movement in normal flight. The landing gear is the understructure that supports the weight of the craft when on the ground or on the water and that reduces the shock on landing.

Airship or **Dirigible**. Aircraft consisting of a cigar-shaped balloon that carries a propulsion system (propellers), a steering mechanism, and accomodations for passengers, crew, and cargo. The balloon section is filled with a lighter-than-air gas—either helium, which is nonflammable, or hydrogen—to give the airship its lift. The balloon maintains its form by the internal gas pressure in the non-rigid (blimp) and semi-rigid types or airships; the latter in addition has a rigid keel. The rigid type maintains its form by having a metal framework that holds its shape regardless of the internal gas pressure; inside the hull are a number of small gas-filled balloons. The first successful power-driven airship was built by the French inventor Henri Giffard in 1852. Count Ferdinand von Zeppelin of Germany invented the first rigid airship, which was completed in 1900. The German airship *Hindenburg* burned at its mooring mast at Lakehurst, N.J., in 1937. No rigid airship survived World War II.

Alabaster. Fine-grained, translucent variety of the mineral Gypsum, pure white or streaked with reddish brown. Its softness makes it easily carved but also easily broken, soiled, and weathered. Quarried in England and Italy, it is used to make statuary and other decorative objects. The Oriental alabaster of ancient Egyptian and Roman tombs is actually Marble, a calcium carbonate, whereas gypsum is a calcium sulfate.

Alanine (2-aminopropanoic acid). A colourless crystalline nonessential amino acid, $CH_3CH(NH_2)COOH$. M.pt. 295-297°C (decomposes).

Albedo. The ratio of the amount of light scattered from a surface to the amount of incident light.

Albino. Animal or plant lacking normal pigmentation. The albino body covering (skin, hair, and feathers) and eyes lack pigment. In humans and other animals albinism is inherited as a recessive trait. Breeding has established albino races in some domestic animals.

Albumin. Member of a class of water-soluble, heat-coagulating Proteins. Albumins are widely distributed in plant and animal tissues, e.g., ovalbumin of egg, lactalbumin of milk, and leucosin of wheat. Some contain carbohydrates. Normally constituting about 55% of the plasma proteins, albumins adhere chemically to various substances in the blood, e.g., Amino Acids, and thus play a role in their transport. Albumins and other blood proteins aid in regulating the distribution of water in the body. Albumins are also used in textile printing, the fixation of dyes, sugar refining, and other important processes.

Alchemy. Ancient art or pseudoscience that sought to turn base metals into gold or silver through the agency of a secret substance known by various names (philosopher's stone, elixir, grand magistry). Emerging in China and Egypt by the 3d cent. B.C., alchemy was cloaked in mysticism and allegory, and in time degenerated into superstitution. Revived (8th cent.) in Alexandria by the Arabs, it reached W. Europe by the Middle Ages. In the 15th-17th cent. experimentation again fell into disrepute, but the base had been laid for modern Chemistry, which has in fact accomplished the transmutation of elements.

Alcohol. 1. Any of a class of organic compounds with the general formula R—Oh, where R is an alkyl group made up of carbon and hydrogen and—OH is one or more hydroxyl groups, each made up of one atom of oxygen and one of hydrogen. Although the term *alcohol* ordinarily refers to Ethanol, the alcohol in alcoholic beverages, the class of alcohols also includes Methanol and the amyl, butyl, and propyl alcohols, all with one hydroxyl group; the glycols, with two hydroxyl groups; and glycerol, with three. Many of the characteristic properties and reactions of alcohols are due to the polarity, or unequal distribution, of electric charges in the C-O-H portion of the molecule.

2. Any of a class of organic compounds with general formula ROH, where R is a hydrocarbon group or substituted hydrocarbon group. Alcohols are distinguished from phenols, in which the hydroxyl group is attached to a carbon atom in an aromatic ring. Alcohols are classified into *primary alcohols,* with general formula RCH_2OH, *secondary alcohols* with formula RR CHOH, and *tertiary alcohols,* with formula RR'-R" OH. They are also classified as monohydric, dihydric, trihydric, etc., according to the number of hydroxyl groups they contain.

Alcohols react with electropositive metals to form alkoxides: $2ROH + 2M = 2MOR + H_2$. They also react with acids to yield esters and water. With oxidizing agents, primary alcohols give aldehydes: $RCH_2OH + O = RCOH + H_2O$; secondary alcohols give ketones: $RCH(OH)R' + O = RCOR' + H_2O$. Concentrated sulphuric acid and other dehydrating agents produce alkenes: $RCH_2CH_2OH - H_2O = RCH{:}CH_2$.

Alcoholics Anonymous (AA). Worldwide organization dedicated to the curing of alcoholics; est. 1935 by two former alcoholics. The organization, which functions through local groups, is based on a philosophy of life that has enabled countless numbers of people to recover from alcoholism. In 1981 there were over 40,000 groups and 1 million members worldwide. *Al-anon*, for spouses, relatives, and friends of alcoholics, and *Al-Ateen,* for their adolescent children, function similarly.

Alcoholism. Chronic illness characterized by the habitual consumption of alcohol to a degree that interferes with physical or mental health, or with normal social or occupational behaviour. A widespread health problem, it produces both physical and psychological addiction. Alcohol is a central nervous system Depressant that reduces anxiety, inhibition, and feelings of guilt; lowers alertness; impairs perception, judgment, and muscular coordination; and, in high doses, can cause unconsciousness and even death. Long-term alcoholism damages the brain, liver, heart, and other organs. Symptoms of alcohol withdrawal can range from a simple hangover to severe delirium tremens (a conditions characterized by deliriousness, violent trembling, hallucinations, and seizures). Treatment includes use of disulfiram (Antabuse), a drug that produces discomfort if alcohol is consumed; anti-anxiety drugs to suppress withdrawal symptoms; psychological counseling; and support from groups such as Alcoholics Anonymous.

Alcohol Thermometer. A type of thermometer consisting of a small glass bulb containing alcohol, connected to a fine sealed capillary tube. The temperature is measured by the expansion of the alcohol, which is usually coloured with red dye. Alcohol has a lower freezing point than mercury and the instrument is useful at lower temperatures.

Aldehyde. Any of a class of organic compounds containing the group -CHO (the *aldehyde group*), consisting of a carbonyl group bound to a hydrogen atom. Aldehydes are produced by oxidizing alcohols and are further oxidized to carboxylic acids: RCO.H + O = RCO.OH. Aldehydes are reducing agents and give a positive reaction with Schiff's reagent, Tollen's reagent, and Fehling's solution. They also have a variety of addition and condensation reactions.

Alder. Deciduous tree or shrub (genus *Alnus*) of the Birch family, widely distributed, especially in mountainous, moist areas of the north temperate zone and in the Andes. The bark of the black alder (*A. glutinosa*), once used medicinally, is still used for dyes and tanning. The red alder (*A. rubra*) is the most important hardwood on the Pacific coast of North America.

Aldol. 1. Any organic compound containing a hydroxyl group and an aldehyde group bound to adjacent carbon atoms, i.e. containing the *aldol group,* -CH(OH)CH(CHO)-.

2. *See* acetal dol.

Aldose. A Sugar that is an aldehyde in its straight-chain form.

Alfalfa or **Lucern.** Perennial plant (*Medicago sativa*) of the Pulse family, probably native to Persia and now widely cultivated. It is an important pasture and hay plant. Alfalfa is valued for its high yield of protein, its effectiveness in weed control, its role in crop rotation and nitrogen fixation, and as a source of chlorophyll and carotene.

Alfven waves. Magnetohydrodynamic waves propagated through a plasma under certain conditions. Alfven waves travel in the direction of the magnetic field, and the particles of plasma oscillate in a plane perpendicular to this direction. The motion of particles is communicated both by collisions and by interaction with electric and magnetic fields. [After Hannes Alfven (b. 1908), Swedish physicist].

Algae. Primitive plants that contain Chlorophyll and carry on Photosynthesis but lack true roots, stems, and leaves. They are the chief aquatic plant life both in the sea and fresh water; nearly all Seaweeds are marine algae. Algae occur as microscopic single cells (e.g., Diatoms) and more complex forms of many cells grouped in spherical colonies (e.g., *Volvox*), in ribbonlike filaments (e.g., *Spirogyra*), and in giant forms (e.g., the marine kelps). The cells of colonies are generally similar, but some are specialized for reproduction and other functions. The blue-green and green algae include most of the freshwater forms, such as pond scum, a green slime found in stagnant water. Brown and red algae are more complex-chiefly marine forms whose green chlorophyll is masked by the presence of other pigments. Algae are primary food producers in the food chain and also provide oxygen for aquatic life.

Algebra. Branch of Mathematics concerned with operations on sets of numbers or other elements that are often represented by symbols. In elementary algebra, letters are used to stand for numbers, e.g., in the Polynomial equation $ax^2 + bx + c = 0$, the letters, *a,b,* and *c* are called the coefficients of the Equation and stand for fixed numbers, or constants. The letter x stands for an unknown number, or variable, whose value depends on the values of *a,b,* and *c* and may be determined by solving the equation. Much of classical algebra is concerned with finding solutions to equations or systems of equations, i.e., finding the Roots, or values of the unknowns, that upon substitution into the original equation will make it a numerical identity. Algebra is a generalization of arithmetic and gains much of its power from dealing symbolically with elements and operations (chiefly addition and multiplication) and relationships (such as equality) connecting the elements. Thus $a + a = 2a$ and $a + b = b + a$ no matter what numbers a and b represent.

Algebraic. Of a relating to algebra or the rules of algebra. The *algebraic sum* of a set of numbers is the result of their addition with due regard to sign; for example the algebraic sum of 25 and -30 is -5 (not 55).

Algin. *See* Alginic Acid.

Alginate. *See* Alginic Acid.

Alginic Acid. A polymeric gelatinous substance extracted from certain seaweeds. Alginic acid has a chain similar to that in carbohydrates with carboxyl groups attached. Its salts, the *alginates,* form viscous gelatinous solutions. Sodium alginate (*algin*) is used as an emulsifier and as a thickener in some foods. Calcium alginate can be spun into fibres.

Algorithm. A procedure or formula applied mechanically in order to carry out a computation. For example, *Euclid's algorithm* is a mechanical method for calculating the highest common factor of two large numbers, by dividing the larger by the smaller, then dividing the smaller by the remainder, then dividing the first remainder by the second, and repeating this procedure until one division is exact. When this occurs, the last divisor is the highest common factor.

Alicyclic. Denoting a chemical compound that is both aliphatic and cyclic. Cyclohexane is a common example.

Aliphatic. Denoting an organic compound that is not aromatic: i.e. does not have the characteristic chemical behaviour of benzene. Aliphatic compounds include the alkanes, alkenes, alkynes, cycloalkanes, and their derivatives.

Alizarin (1,2-dihydroxyanthraquinone). An orange-red crystalline solid, $C_6H_4(CO)_2C_6H_2(OH)_2$, used as a dyestuff. It was formerly extracted from madder roots but is now synthesized. M.pt. 289°C; b.pt. 430°C.

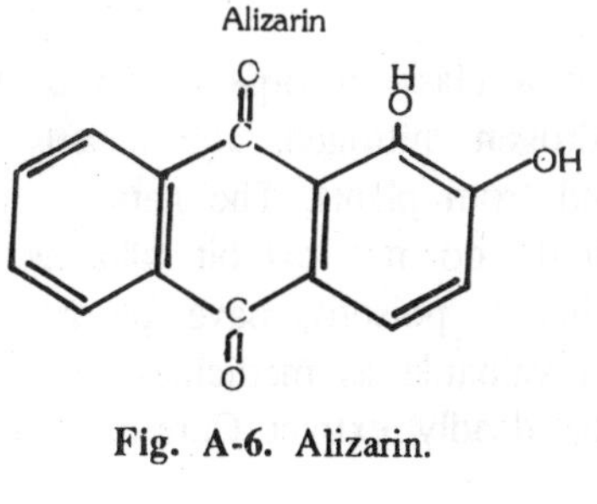

Alizarin

Fig. A-6. Alizarin.

Alkali. Hydroxide of an Alkali Metal. Alkalies are soluble in water and form strongly basic solutions. They neutralize acids, forming salts and water. Strong alkalies (e.g., those of sodium or potassium) are called caustic alkalies. The term *alkali* is sometimes applied to sodium or potassium carbonate or to the hydroxide of an Alkaline-Earth Metal.

Alkali Metals. Elements in group la of the Periodic Table. In order of increasing atomic number, they are Lithium, sodium, Potassium, Rubidium, Cesium, and Fracium. They are softer than other metals, and have lower melting points and densities. All react violently with water, releasing hydrogen and forming hydroxides. They tarnish rapidly, even in dry air. They never occur uncombined in nature.

Alkalimetry. Determination of the amounts of alkalis in solution by titration.

Alkaline-earth metals. The metallic elements beryllium, magnesium, calcium, strontium, and barium, belonging to group IIA of the periodic table. They are silvery-white metals, which are divalent and electropositive and are not quite as reactive as the alkali metals. Like the alkali metals they form basic oxides and hydroxides and crystalline ionic salts. Strictly speaking the *alkaline earths* are the oxides of these metals, although the term is often used for the metals themselves.

Alkaline earths. Oxides of the Alkaline-Earth Metals. They are not readily soluble in water and form solutions less basic than those of Alkalies.

Alkaloid. Any of a class of organic compounds composed of carbon, hydrogen, nitrogen, and usually oxygen, that are often derived from plants. The name means alkalilike, but some alkaloids do not exhibit alkaline properties. many alkaloids, though poisons, have physiological effects that render them valuable as mediçines. For example, curarine, found in the deadly extract Curare, is a powerful muscle relaxant; atropine is used to dilate the pupils of the eye; and physostigmine is a specific for certain muscular diseases. Narcotic alkaloids used in medicine include Morphine and Codeine for paint relief and Cocaine as a local anesthetic. Other common alkaloids include Caffeine, LSD, quinine, Serotonin, Strychnine, and nicotine.

Alkane (paraffin). Any of the saturated hydrocarbons with general formula C_nH_{2n+2}. The alkanes form a homologous series starting with methane, CH_4. The first four alkanes are gases and higher members of the series are liquids or colourless waxes. Alkanes tend to be rather unreactive: they can be converted into alkyl halides by substitution.

Alkanization. Hydrogenation of an unsaturated hydrocarbon, such as an alkene, to produce an alkane.

Alkene. Any one of a class of hydrocarbons characterized by the presence of double bonds between carbon atoms. The simplest example is ethylene. Alkenes can be prepared by treatment of an alkyl halide with alcoholic potassium hydroxide: C_2H_5Cl+ KOH = CH_2:CH_2 + KCl + H_2O. Other common methods are the dehydration of alcohols, dehalogenation of vicinal dihalides, and reduction of alkynes. The main type of reaction of alkenes are addition reactions. For example, under acid conditions: CH_2:CH_2 + H_2O = CH_3CH_2OH. Under some conditions substitution can also occur : CH_3CH_2: CH_2 + Cl_2 = CH_2ClCH_2:CH_2. Alkenes also react with ozone to form an unstable intermediate which breaks down to give aldehydes or ketones: RCH:CHR' + O_3 = RCOH + R'COH. Alkenes with one double bond form a homologous series with general formula C_nH_{2n} , called the *ethylene series*.

Alkyd resin. *See* Polyester resin.

Alkyl. An organometallic compound in which a metal atom is directly bound to which a metal atom is directly bound to an alkyl group. Tetraethyl lead, $Pb(C_2H_5)_4$, is a common example.

The aluminium alkyls, with general formula $Al(C_nH_{2n+1})_3$, are also commercially important, being used in Ziegler-Natta catalysts. Alkyls can be made by direct reaction of a metal with an alkyl halide.

Alkylation. The introduction of an alkyl group into an organic compound in a chemical reaction. Usually alkylations involve addition of an alkyl group to an alkene, or substitution of an alkyl group into an aromatic ring by a Friedel-Crafts reaction.

Alkylbenzene. Any hydrocarbon containing an aromatic group bound to an alkyl group, and having the chemical properties of both.

Alkyl group. A univalent organic group derived by removing one hydrogen atom from an aliphatic hydrocarbon, usually an alkane. An example is the methyl group, CH_{3-}.

Alkyne. Any one of a class of hydrocarbons characterized by the presence of triple bonds between carbon atoms.

Allene. A colourless gaseous hydrocarbon, CH_2:C:CH_2.

Alligator, aquatic Reptile (genus *alligator*) in the same order as the Crocodile. The American alligator (*A. mississipiensis*) is found in swamps and streams from North Carolina to Florida and along the Gulf Coast. The young are brown or black with yellow bands; adults are solid black. Males commonly reach a length of 9 ft (2.7 m) and a weight of 250 lb (110 kg). The alligator is protected by law.

Allotropy. The occurrence of certain chemical elements in two or more forms; the forms are called allotropes. Allotropes generally differ in physical properties, such as color and hardness; they may also differ in molecular structure or in chemical activity but are usually alike in most chemical properties. Diamond and Graphite are two allotropes of the element Carbon.

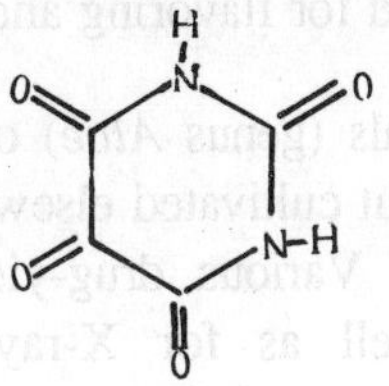

Fig. A-7. Alloxan.

Alloxan. A white crystalline heterocyclic compound, $C_4H_2O_4N_2$. M.pt. 170°C (decomposes).

Alloy. Substance with metallic properties consisting of a Metal fused with one or more metals or nonmetals. An alloy may be a homogeneous solid solution, a heterogeneous mixture of tiny crystals, a true chemical compound, or a mixture of these. Alloys generally have properties different from those of their constituent elements and are used more extensively than pure metals. Alloys of Iron and Carbon are among the most widely used and include Cast Iron and Steel Brass and Bronze are important alloys of Copper. Because pure Gold and Silver are too soft for many uses, they are often alloyed, either with each other or with other metals, e.g., copper or Platinum. Amalgams arc alloys that contain Mercury. Other alloys include Britannia Metal, Duralumin, and Solder.

Allyl alcohol (prop-2-en-1-ol). A colourless poisonous liquid, CH_2:$CHCH_2OH$, with a pungent mustard-like odour, used in the manufacture of plasticizers and synthetic resins. M.pt. -129°C; b.pt. 97°C; r.d. 0.85.

Allyl Group. The organic group CH_2:-$CHCH_2$-.

Almond. Name for a small tree (*Prunus amygdalus*) of the Rose family and for the nutlike edible seed of its drupe fruit. Almonds are now cultivated principally in the Orient, Italy, Spain, and California. Almond fruit is fleshless; otherwise, the tree closely resembles the peach tree. Almond oil, pressed from the nut, is used for flavoring and in soaps and cosmetics.

Aloe. Succulent perennials (genus *Aloe*) of the Lily family native chiefly to Africa, but cultivated elsewhere. The leaves contain aloin, a purgative. Various drug-yielding species are used medicinally, as well as for X-ray-burn treatment, insect repellent, and transparent pigment. Biblical aloes is unrelated. American and false aloe are agaves of the Amaryllis family.

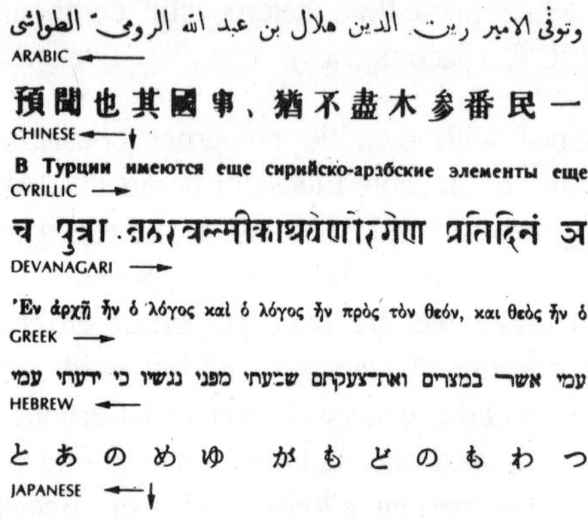

Fig. A-8. Alphabet: Examples of letters in various alphabets (arrows indicate the direction of reading).

Alphabet. System of Writing, theoretically having a one-for-one relation between character (or letter) and phoneme. Few alphabets have achieved an ideal exactness. A system of writing is called a syllabary when one character represents a syllable rather than a phoneme, e.g., the kana used in Japanese. The precursors of the alphabet were the iconographic and ideographic writing of ancient man, such as Cuneiform and the Hieroglyphic writing of the Egyptians. The alphabet of modern Western Europe is the roman alphabet. Russian, Serbian, Bulgarian, and many languages of the USSR are written in the Cyrillic alphabet, an augmented Greek alphabet. Greek, Hebrew, and Arabic all have their own alphabets.

The most important writing of India is the Devanagari, an alphabet with Syllabic features. The Roman alphabet is derived from the Greeks, who had imitated the Phoenician alphabet. The exact steps are unknown, but the Phoenician, Hebrew, Arabic, and Devanagari systems are based ultimately on signs of Egyptian hieroglyphic writing. Two European alphabets of the late Roman era were the Runes and the ogham. An exotic modern system is the Cherokee syllabary of Sequoyah.

Alpha decay. A type of radioactive decay in which an unstable nucleus ejects on alpha particle. The product has a mass number that has decreased by 4 and an atomic number that has decreased by 2. Radium-226, for example, with an atomic number of 88, decays to give radon-222, which has an atomic number of 86. The radiation emitted during alpha decay follows the Geiger-Nuttall law.

Alpha Iron. The allotropic form of iron that is stable below 906°C. It has a body-centred cubic crystal structure and is ferromagnetic up to its Curie temperature (768°C).

Alphanumeric or **Alphameric.** The alphabetic, numeric, and special characters—such as mathematical symbols and punctuation marks—used in Computer input and output.

Alpha Particle (α-particle). A helium-atom nucleus, containing two protons and two neutrons. Alpha particles are emitted in streams (*alpha rays*) by radioactive decay.

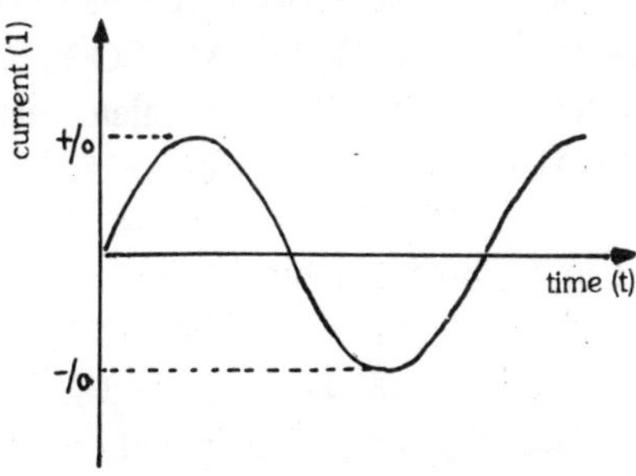

Fig. A-9. Alternating current.

Alternating Current (A.C.). An electric current that varies in strength, periodically reversing its direction. The commonest practical form is a sinusoidal current, in which the value of the current follows the equation $I = I_o \sin 2\pi ft$. I_o is the peak value of the current, f is the frequency, and t is the time.

Alternating Series. As infinite series whose elements are alternately positive and negative, e.g. $1 - {}^1/_2 + {}^1/_2 - {}^1/_6 + {}^1/_{12} \cdot \ldots$

Such a series converges if the moduli of the successive terms decrease and tend to zero.

Alternator. A device for producing an alternating current. See generator.

Altimeter. Device for measuring altitude. The most common type, used in airplanes and balloons and by mountain climbers, consists of an aneroid Barometer, calibrated so that the drop in atmospheric pressure indicates the linear elevation. The radio altimeter indicates the actual altitude over the earth's surface by measuring the time it takes for a radio signal to travel to earth and back.

Altitude. 1. A celestial coordinate that determines the angular distance of a star, planet, etc., above the horizon.

2. The height of a geometric figure or solid measured perpendicular to the base.

Alum. Any of a class of isomorphous hydrated crystalline double salts having the formula, $M_2SO_4.M'_2(SO_4)_3.24H_2O$, where M is a monovalent metal and M' a trivalent metal. The term is often applied to the double sulphate of potassium and aluminium, K_2SO_4, Al_2-$(SO_4)_3.24H_2O$, known as *potash alum.* The potassium can be replaced by other monovalent ions as in *ammonium alum,* $(NH_4)_2SO_4.Al_2(SO_4)_3 24H_2O$. Similarly the term alum is used for double salts in which the aluminium has been replaced by a trivalent metal, as in *chrome alum,* $K_2SO_4.Cr_2(SO_4)_3.24H_2O$, or *ammonium ferric alum,*

$(NH_4)_2SO_4.Fe_2(SO_4)_3.24\text{-}H_2O$. Alums are prepared by crystallizing solutions of the sulphates mixed in the correct proportions. In strict chemical usage they are named according to the rules for double salts: aluminium potassium sulphate, etc.

Alumina or **Aluminum Oxide**. Chemical compound (Al_2O_3) that is widely distributed in nature and occurs combined with silica and other minerals in clays, feldspars, and micas, and in almost pure form in Corundum. It is the major component of Bauxite and is used in the production of Aluminum metal. Alumina is also used as an Abrasive, in ceramics, in pigments, and in the manufacture of chemicals.

Aluminium. Symbol: Al. A silvery ductile metallic element obtained from bauxite; the most abundant metal in the earth's crust. It is produced by converting the bauxite to aluminium oxide by addition of caustic soda and heating the precipitated aluminium hydroxide. a fused mixture of the oxide with cryolite is electrolysed to obtain the aluminium. It is extensively used in a wide range of light aluminium alloys, which also contain manganese, nickel, copper, zinc, and other metals and are used in aircraft, ships, machinery, cooking utensils, etc. It is also a good conductor of electricity. Chemically, aluminium is a reactive metal, forming a film of transparent oxide in air, which protects against further oxidation. It also reacts with other nonmetals. Aluminium is more typically metallic than boron, the first member of its group in the periodic table, and forms ionic crystalline salts such as aluminium sulphate. The halides, however, are volatile compounds, dimeric in the vapour, and seem to be covalently bonded indicating a degree of non-metallic behaviour. A.N. 13; A.W. 26.9815; m.pt. 659.7°C; by b.pt. 1800°C; r.d. 2.6; valency 3.

Aluminium Hydroxide. A white odourless tasteless powder, also obtainable as a gel, $Al(OH)_3$. A hydrated amphoteric oxide, it is prepared by adding alkali to a solution of an aluminium salt. It is used as a mordant for dyes and as a gastric antacid.

Aluminium Oxide (alumina). A white powder, Al_2O_3. An amphoteric oxide that occurs naturally as corundum. Emery, ruby, and sapphire are impure crystalline forms. Aluminium oxide is obtained commercially from bauxite and its principal use is in the manufacture of aluminium. It is also used in abrasives, ceramics, and chromatography. M.pt. 2030°C; r.d. 3.4-4.0.

Alunite. A white or pink mineral consisting of a basic sulphate of aluminium and potassium, $K_2SO_4.Al_2(SO_4)_3.4Al\text{-}(OH)_3$, used as a source of potash alum.

Alyssum. Chiefly annual and perennial herb (genus *Alyssum*) of the Mustard family, native to the Mediterranean. Some species, with masses of yellow or white flowers, are cultivated as rock-garden and border ornamentals. The annual sweet alyssum has fragrant white or lilac blossoms. Once called madwort or healbite, alyssum was thought to cure rabies.

Alzheimer's disease. Degenerative disease of the brain cell producing loss of memory and general intellectual impairment. It usually affects people in their forties, fifties, and sixties. As the disease progresses, a variety of symptoms may become apparent, including confusion, irritability, and restlessness, as well as disorientation and impaired judgment and concentration. The cause of the disease is unknown, and there is no known treatment, although tranquilizers are sometimes used to reduce agitation and unpredictable behavior.

Amalgam. An alloy of mercury with one or more other metals. Amalgams are either liquid or solid, some being simple mixtures and others containing definite compounds (such as Hg_2Na). Most metals will form amalgams: iron and platinum are notable exceptions.

Amaranth. Common name for the family Amaranthaceae (also called the pigweed family), herbs, trees, and vines found mainly in warm regions of the Americas and Africa. The genus *Amaranthus* includes several species called amaranth, characterized by a lasting red pigment in the stems and leaves; common weeds, e.g., the green amaranth (*A. retroflexus)*; and various plants known as Tumbleweed and pigweed. Some species have long been used as potherbs and cereals in the Old and New World. The globe amaranth, or Bachelor's Button (genus *Gomphrenia*), and the cockscomb (genus *Celosia*) are tropical annuals that are dried and used in everlasting bouquets.

Fig. A-10. *Green amaranth,* Amaranthus retroflexus.

Amaryllis. Common name for some members of the Amaryllidaceae, a family of mostly perennial plants with flat, narrow leaves and lilylike flowers borne on separate, leafless stalks. Widely distributed, they are found especially in tropical and subtropical lowlands. Ornamentals of the family are mistaken for plants of the Lily family, which differ in the position of the ovary. The family includes the showy-blossomed true amaryllis, or belladonna lily (*Amaryllis belladonna*; Narcissus; and the snowdrops (genus *Galanthus),* whose small, early-blooming flowers are symbolic of consolation and promise. The genus *Agave,* the tropical american counterpart of the African genus Aloe, contains the most economically important plants in the family. Different agaves provide soap, food, beverages, and fibe..

Amatol. An explosive consisting of a mixture of ammonium nitrate and TNT.

Amber. Yellow to brown fossil Resin exuded by coniferous trees now extinct; the best amber is transparent. Highly polished amber is used to make small decorative objects, e.g., beads and amulets. When rubbed with a cloth, amber becomes charged with static electricity. Bubbles of air, leaves, bits of wood, or insects, sometimes of extinct species, are often found trapped in amber. The chief source of the world's amber is the Baltic coast of Germany.

Ambergris. A waxlike substance that is found in the intestine of the sperm whale. It is found floating on tropical seas or cast up on the shore in yellow, gray, black, or variegated masses. Greatly valued from earliest times, ambergris is now used as a fixative in perfumes.

Americium. Symbol: Am. A silvery-white transuranic actinide element made by neutron bombardment of plutonium. The most stable isotope. ^{243}Am. has a half-life of 8800 years. A.N. 95; r.d. 11.7; valency 3,4,5, or 6.

Americium. (Am), synthetic radioactive element, discovered by Glenn Seaborg and colleagues in 1944 by neutron bombardment of plutonium. The silver-whgite metal is in the Actinide Series. Half-lives of the many isotopes range from 1.3 hr to over 7,000 years.

Amethyst. Most highly valued variety of Quartz, violet to purple in color, used as a Gem. Amethyst is found in Brazil, Uruguay, Sri Lanka, Siberia, and North America. It has superstitious associations, being regarded as a love charm, a sleeping aid, and a guard against thieves and drunkenness.

Amine. Any of a class of organic compounds derived by replacing one or more of the hydrogen atoms of ammonia with an alkyl or aryl group. If one hydrogen is replaced, a *primary* amine results having a formula RNH_2. *Secondary* amines have the general formula RR'NH and *tertiary* amines have the formula RR'R"N. Methods of preparing amines include reduction of nitro compound, nitriles, and amides. They are basic substances, forming quaternary salts, and also react with nitrous acid to give diazo compounds.

Amino Acid. Any of a class of organic compounds that contain both an amino group and a carboxyl group in their molecules. Glycine, G_2NCH_2COOH, is the simplest example. Amino acids are the units present in peptides and proteins. Most proteins contain amounts of 23 amino acids. *Essential* amino acids are those that have to be present in the diet of an animal for conversion into protein: *nonessential* amino acids can be synthesized in the body. There are eight essential amino acids in the diet of man.

Amino Acid. Some 22 amino acids are commonly found in animals and more than 100 less common forms are found in nature, chiefly in plants. When the carboxyl carbon atom of one amino acid binds to the nitrogen of another with the release of a water molecule, a linkage called a peptide bond is formed. Chains of amino acids, joined head-to-tail in this manner, are synthesized by living systems and are called polypeptides (up to about 50 amino acids) and Proteins (over 50 amino acids). The process of digestion releases individual amino acids from food protein by cleaving the peptide bonds.

Aminoethane. *See* Ethylamine.

Amino Group. The group $-NH_2$, as present in amines.

Aminoplastic. *See* Amino resin.

Amino Resin (Aminoplastic). Any of a class of synthetic resins made by copolymerizing an aldehyde, usually formaldehyde, with an amine. The most important amino resins are melamine and urea-formaldehyde.

Ammeter. An instrument for measuring an electric current. The moving-coil instrument is fairly sensitive but can be used only for direct current. Alternating currents are measured by the moving-iron instrument, which is less sensitive and has a nonlinear scale. Both types have very low resistances.

At high frequencies *hot-wire ammeters* are used. The current to be measured passes along a thin resistance wire, causing a rise in temperature. This is detected by a thermocouple or by the resulting expansion of the wire.

Ammine. An inorganic metal complex containing coordinated ammonia molecules, as in copper ammine, $[Cu\text{-}(NH_3)_4]^{2+}$.

Ammonal. An explosive consisting of a mixture of ammonium nitrate and aluminum powder.

Ammonia. A colourless pungent toxic gas, NH_3, manufactured by methods that have evolved from the Haber process of direct catalytic combination of nitrogen and hydrogen. It is used in the manufacture of fertilizers, nitric acid, explosives, and synthetic fibres. Ammonia is often used in aqueous solution, which is alkaline as a result of hydroxide ions (OH) being produced with ammonium ions (NH_4^+). The compound NH_4OH (*ammonium hydroxide*) is unstable and cannot be isolated. M.pt. -78°C; b.pt. -33.5°C.

Ammoniacal. Denoting the presence of ammonia. The term is usually used to describe solutions in aqueous ammonia.

Ammonia Soda Process. *See* Solvay process.

Ammonium Alum. *See* Alum.

Ammonium Chloride (Sal ammoniac). A white crystalline solid, NH_4Cl, manufactured as a by-product in the Solvay process.

It is used in batteries and as a pickling agent in zinc coating and tinning. Sublimes without melting; r.d. 1.54.

Ammonium Hydroxide. *See* Ammonia.

Ammonium ion. The monovalent ion NH_4^+ produced by protonation of ammonia. Ammonium salts contain these ions: ammonium chloride, for example, is NH_4^+ Cl^-

Amnesia. Condition characterized by loss of memory for long or short intervals. It may be caused by physical injury, Shock, senility, or severe illness. In other cases, a painful experience and everything remindful of it, including the individual's identity, is unconsciously repressed. Attempts to cure this type of amnesia include efforts to establish associations with the past through suggestion and Hypnosis.

Amniocentesis. Diagnostic procedure in which a sample of the amniotic fluid surrounding the fetus is removed from the uterus with a fine needle inserted through the abdomen of the pregnant woman (See Pregnancy). Fetal cells in the fluid can be grown in the laboratory and studied to detect the presence of certain genetic disorders (e.g., Down's Syndrome, Tay-Sachs Disease) or physical abnormalities. Generally recommended when there is family history of genetic disorders or when the woman is over age 35, the procedure is usually carried out around the 14th or 15th week of pregnancy, when there is sufficient amniotic fluid and Abortion is still an option.

Amoeba or Ameba. One-celled organism (class Sarcodina) in the phylum Protozoa. Amebas constantly change their body shape as they form temporary extensions called pseudopods, or false feet, used for feeding and locomotion. Most amebas range from 5 to 20 microns in diameter. They engulf their prey (diatoms, algae, bacteria) with their pseudopods, forming vacuoles in which food is digested by Enzymes. Reproduction is usually by binary fission (splitting) to produce two daughter amebas, the nucleus dividing by Mitosis; some also reproduce sexually. Amebas live in fresh and marine waters and the upper layers of soil. Many are Parasites of aquatic and terrestrial animals, and some cause disease, e.g., amebic dysentery.

Amorphous. Denoting a solid substance with no regular crystalline structure. The term specifically refers to the lack of order in the arrangement of atoms in the solid-many apparently amorphous substances are composed of minute crystals and are in fact crystalline. True amorphous solids are uncommon: examples are soot (carbon), flowers of sulphur, and glass.

Amount of Substance. Symbol: n. A measure of quantity proportional to the number of particles of substance present. The amount of substance is not the same as the mass. It is always used with reference to a specified type of particle. Thus the amount of substance of an element is proportional to the number of atoms present; the amount of substance of light is proportional to the number of photons, etc. The constant of proportionality is the Avogadro Constant. Amount of substance is always measured in moles.

Ampere. Symbol: A. The basic SI unit of electric current, equal to the current that, when passed through two parallel conductors of infinite length and negligible circular cross section placed one metre apart in a vacuum, produces a force of 2×10^{-7} newton per sq. metre between them. Formerly, this was the *absolute ampere,* taken to be one tenth of the CGS-electromagnetic unit of current. The international unit of current, the *international ampere,* was defined in 1893 as the current that, when passed through a solution of silver nitrate under specified conditions, deposits silver at a rate of 0.001 118 gram per second. In 1948 the international ampere was redefined as 0.999 850 A. [After Andre-Marie Ampere (1775-1836), French physicist.]

Ampere, Andre Marie. 1775-1836, French physicist, mathematician, and natural philosopher. He extended the work of Hans Oersted on the relationship of electricity and magnetism, formulated Ampere's law describing the contribution of a current element to magnetic induction, and invented the static needle. The basic unit of electric current, the Ampere, is named for him.

Ampere-hour. A unit of quantity of electricity (charge) equal to the amount flowing in one hour as the result of a steady current of one ampere. It is equal to 3600 coulombs.

Ampere's law. The formula $dB = (\mu_o I \sin \theta . dl)/4\pi r^2$, giving the elemental magnetic flux density dB produced by a current I at a point a distance r from a conductor element dl. A line from the point to the conductor makes an angle θ with the current direction and μ_o is the magnetic constant.

Ampere-turn. Symbol: A or AT, A unit of magnetomotive force equal to the magnetomotive force produced by a current of one ampere flowing through one turn of a coil.

Amphetamine. Any of a class of powerful drugs that act as stimulants (See Drugs) on the central nervous system. Popularly known as "bennies", "speed", or "uppers", amphetamines enhance mental alertness and the ability to concentrate; cause wakefulness, talkativeness, and euphoria; and temporarily reverse the effects of fatigue. They have been used to treat obesity, narcolepsy and minimal brain dysfunction. Amphetamines can produce insomnia, hyperactivity, and irritability, as well as such severe systemic disorders as cardiac irregularities, elevated blood pressure, and gastric disturbances. The drugs are addictive and easily abused; addiction can result in psychosis or death from overexhaustion or cardiac arrest.

Amphibian. Cold-blooded animal of the class Amphibia, the most primitive of terrestrial Vertebrates. Unlike Reptiles, the amphibians, which include Frogs and Toads, Salamanders and newts, and the limbless Caecilians, have moist skins without scales or with small, hidden scales. Most amphibians deposit their eggs in water or a moist, protected place. The young undergo Metamorphosis from aquatic, water-breathing, limbless Tadpoles to terrestrial or partly terrestrial, air-breathing, four-legged adults.

Amphitheater. Open structure for the exhibition of gladiatorial contests and spectacles, built in cities throughout the Roman Empire, e.g., at Rome, Arles (France), and Cirencester (England). The typical amphitheater was elliptical, with seats rising around a central arena; quarters for the gladiators and animals were under the arena. The word is now used for various quite unrelated structures.

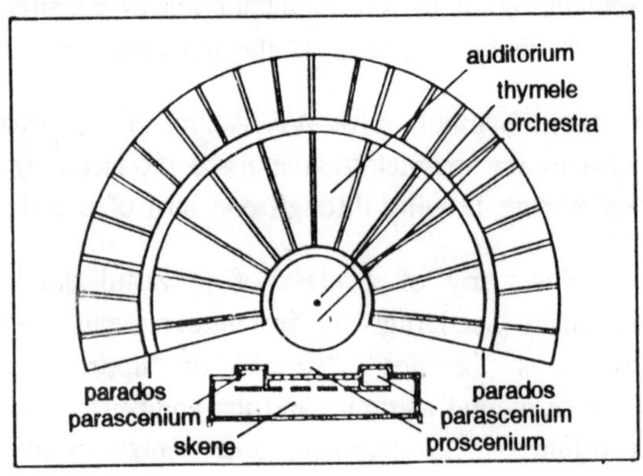

Fig. A.11. *Amphitheater.*

Ampholyte ion. *See* Zwitterion.

Amphoteric. Capable of reacting both as an acidic substance and a basic substance. For example, with strong acids aluminium hydroxide acts as a base: $Al(OH)_3 + 3HCl = AlCl_3 + 3H_2O$. With strong bases it acts as an acid to form aluminates : $NaOH + Al(OH)_3 = NaAlO_2 + 2H_2O$.

Amphoterism. In Chemistry, the property of certain compounds of acting either as acids or as bases (See Acids and Bases), depending on the reaction in which they are involved. Many hydroxide compounds and organic molecules that contain both acidic (e.g., carboxyl) and basic (e.g., amino) Functional Groups are usually amphoteric.

Amplifier. Device in which a varying input signal controls a flow of energy to produce an output signal that varies in the same way but has a larger amplitude; the input signal may be a current, a voltage, a mechanical motion, or any other signal, and the output signals is usually of the same nature. The ratio of the output voltage to the input voltage is called the voltage gain. The most common types of amplifiers are electronic and have Transistors as their principle components. In most cases today, transistors are incorporated into Integrated-Circuit chips. Most amplifiers include more than one transistor. Transistor amplifiers are used in Radio and Television transmitters and receivers, stereophonic Record Players, and intercoms.

Amplitude. 1. The maximum departure of an oscillating system from its equilibrium value. The amplitude of an alternating current, for instance, is the peak value of the current. The amplitude of a pendulum is the maximum distance is swings from its central position.

2. (Of a complex number) The angle subtended with the x axis by the line joining the representation of the given point on the Argand diagram with the origin See also argument. Compare modulus).

Amplitude Modulation. *See* Modulation.

Amu. *See* Atomic Mass unit.

Amyl Acetate. A colourless flammable liquid with a characteristic odour of pear drops. It is a mixture of esters with the formula $CH_3COOC_5H_{11}$, prepared from amyl alcohol and used chiefly as a solvent for lacquers and paints. The pure isomers can also be obtained.

Amyl Alcohol. Any of eight aliphatic alcohols that are structural isomers with the formula $C_5H_{11}OH$. They are distinguished as n-*amyl alcohol,* sec-*amyl alcohol,* etc. The amyl alcohols are now more usually named according to the main chain of carbon atoms; *pentan-1-ol, pentan-2-ol,* etc. The term is still used for commercial mixtures of alcohols.

Amyl Group. An organic group with the formula $C_5H_{11}-$. Several isomeric forms exist.

Analgesic. A drug used to relieve pain. Analgesics include nonnarcotics, such as Aspirin and other salicylates, which also reduce fever and inflammation; Narcotics, such as Morphine and codeine; and synthetic narcotics with morphinelike action, such as propoxyphene (Darvon) and meperidine (Demerol). Acetaminophen (Tylenol) is often given to individuals sensitive or allergic to salicylates.

Analog Circuit. Electric Circuit in which the output voltage and current values are considered significant over a continuum. Analog circuits may be used for such purposes as amplifying signals corresponding to sound waves.

Analog Computer. A type of computer in which problems are solved by using a mechanism or electric circuit that has the same behaviour as the system investigated. Analog computers are used for solving differential equations or for investigating or simulating complex physical systems. Usually they employ electrical circuits in which variables are represented by currents or voltages input to a network of amplifiers and other components. The network is designed so that the relationship of the output signal to the input signal is given by the equation under investigation. A trivial example would be a single fixed resistance R through which is passed a current I. The potential difference across the resistor is given by IR and in principle the circuit could be used for multiplying an input variable I by a constant factor R. Mechanical devices operate by gears, levers, etc., and variables are represented by distances.

In analog computers the input is continuous rather than discrete.

Analog-to-digital coversion. The process of changing continuously varying data into digital quantities that represent the magnitude of the data at the moment the conversion is made. The most common use is to change analog signals into a form that can be manipulated by digital Computer, as in data communications.

Analysis. Determination of substances or elements present in a sample. *Qualitative analysis* is simply the indentification of a substance or of the constituents of a mixture. *Quantitative analysis* measures the amounts of each substance present. Many different chemical and physical analytical techniques are in use.

Analysis. Branch of Mathematics that uses the concepts and methods of the Calculus. It includes basic calculus; advanced calculus, in which such underlying concepts as that of a Limit are subjected to rigorous examination; differential and integral equations, in which the unknowns are Functions rather than numbers; Vector and tensor analysis; differential geometry; and many other fields.

Analytic Geometry. Branch of Geometry in which points are represented with respect to a coordinate system, such as Cartesian coordinates. Analytic geometry was introduced by Rene Descartes in 1637 and was of fundamental importance in the development of the Calculus by Sir Isaac Newton and G.W. Leibniz in the late 17th cent. Its most common application—the representation of Equations involving two or three variables as curves in two or three Dimensions or surfaces in three dimensions—allows problems in Algebra to be treated geometrically and geometric problems to be treated algebraically. The methods of analytic geometry have been generalized to four or more dimensions and have been combined with other branches of geometry.

Anatomy. Branch of biology concerned with the study of the body structure of plants and animals. Comparative anatomy considers structural similarities and differences of various organisms, forming the basis of Classification. Human anatomy emphasizes individual systems composed of groups of Tissues and Organs.

Andromeda Galaxy. Closest spiral Galaxy (2 million Light-years distant) to our Milky way galaxy. They are similar in shape and composition. The Andromeda galaxy, visible to the naked eye as a faint patch in the constellation Andromeda, is about 120,000 light-years in diameter and contains at least 200 billion stars.

Anemia. Condition resulting from a reduction in Hemoglobin Content or in number of red blood cells (erythrocytes). Although the causes of anemia vary, because of the blood's reduced capacity to carry oxygen all types exhibit similar symptoms—pallor, weakness, dizziness, fatigue, and in severe cases, breathing difficulties and heart abnormalities. Therapy includes identifying and treating the underlying cause.

Anemone or **Windflower.** Wild or cultivated perennial herb (genus *Anemone* of the Buttercup family. Anemones, which contain a poisonous compound (anemonin), were once used to treat fevers, bruises, and freckles; they are often associated with evil and death. The purple-or white-blossomed wood anemone *A. quinquefolia* is also called windflower.

Anesthesia. Loss of sensation, especially that of pain, induced by drugs. General anesthesia, first used in surgical procedures in the 1840s, induces unconsciousness; halothane and sodium pentothal are two widely used general anesthetics. Localized or regional anesthesia is used to desensitize the area around the site of application without loss of consciousness. Regional anesthetics include spinal anesthetics, primarily for lower body surgery; epidural anesthetics for childbirth; and local anesthetics (e.g., novocaine and lidocaine) for dental procedures. Various anesthetics are often used in combination.

Anethole. A white crystalline derivative of anisole, CH_3CH : $CHC_{6H4}OCH_3$, found in some essential oils. It is used in perfumery. M.pt. 22° c; b.pt. 234-7°C; r.d. 0.98.

Angiosperm. Plant in which the ovules, or young seeds, are enclosed within an ovary (that part of a Flower specialised for speed production), in contrast to the Gymnosperms. Also knows as the flowering plants, angiosperms have Leaves, Roots, and Stems and vascular, or conducting tissue. Constituting the plant division Magnoliophyta they are divided into dicotyledons (class Magnoliopsida), which have two seed leaves (cotyledons) and Cambium tissue in the stems, and monocotyledons (class Liliatae), which have one seed leaf and generally lack cambium tissue. The most important plant group economically, angiosperms include all agricultural crops and cereals Grains, all garden flowers, and almost all broad-leaved trees and shrubs.

Angle. In mathematics, figure formed by two straight lines meeting at a point; the point is known as the vertex of the angle, and the lines are its sides. Angles are commonly measured in degree (°) or in radians. If the two sides form a straight line but do not coincide, the angle is a straight angle, measuring 180 or π radians (see pl). If the sides are perpendicular, i.e., if they form half of a straight angle, the angle is a *right angle,* measuring 90° or π/2 radians. An acute angle is greater than 0° but less than 90°; an *abtuse angle* is greater than 90° but less than 180°; and a *reflex angle* is greater than 180°.

Angstrom, Anders Jons. 1814-74, Swedish physicist. Noted for his study of light, especially Spectrum analysis, he mapped the sun's spectrum, discovered hydrogen in the solar atmosphere, and was the first to examine the spectrum of the Aurora borealis. The length unit Angstrom is named for him.

Angstrom (angstrom unit). Symbol: A. A unit of length equal to 10^{-10} metre. It is used for expressing wavelengths of electromagnetic radiation and interatomic distances in molecules and crystals.

Angular frequency. Symbol: ω Frequency measured in radians per unit time, given by 2 π f. where f is the frequency in hertz.

Aangular inomentum. See momentum.

Angular velocity. Symbol: The rate of rotation of a body expressed as the angle turned through per unit time. It is usually measured in radians per second.

Anhydrite. A naturally occurring anhydrous form of calcium sulphate, $CaSO_4$, used in making sulphuric acid.

Anhydrous. Without water. The term is often used to describe salts that have no water of crystallization, as in anhydrous copper sulphate.

Aniline. A colourless oily flammable liquid amine, $C_6H_5NH_2$, prepared by the reduction of nitrobenzene. It is used as the parent substance for many dyes, plastics, and pharmaceuticals. M.pt.-6° C; b.pt. 184° C; r.d. 1.02.

Anilinium ion. The positive ion $[C_6H_5NH_3]^+$, obtained by protonating aniline. Anilinium Salts containing this ion are produced by reaction of aniline with acids, as in the reaction yielding anilinium chloride (aniline hydrochloride): $C_6H_5\ NH_2 + HCl = C_6H_5\ NH_3\ Cl$.

Animal. Any member of the animal kingdom (Animalia) as distinguished from the plant kingdom (plantae). The major difference between animals and plants is their method of obtaining nourishment—animals must secure food as already organized organic substances, while plants manufacture their own food from inorganic sources (usually by Photosynthesis). Most animals have nervous systems, sense organs, and specialized modes of locomotion, and are adapted for securing ingesting, and digesting food. Animals and plants are interdependent: green plants use carbon dioxide and provide oxygen (a by product of photosynthesis) and food; animals (and plants) provide carbon dioxide (through respiration and the decomposition of their dead bodies). The scientific study of animals is called Zoolozy.

Animal charcoal (bone black). A form of carbon obtained by heating bones. It contains large amounts of calcium phosphate. Activated animal charcoal is used as an absorbent for decolourizing, etc.

Animism. Belief that a spirit or force residing in every animate and inanimate object, every dream and idea, gives individuality to each. The related polynesian concept of *mana* holds that the spirit in all things is responsible for the good and evil in the universe.

Anion (negative ion). An ion with a negative charge. Such ions are attracted to the anode during electrolysis. *Compare* cation.

Anise. Annual plant *(Pimpinella anisum)* of the CARROT family, native to the Mediterranean but widely cultivated for its aromatic and medicinal qualities. The seedlike fruits (aniseed) are used in medicinals, perfumes, beverages, and dentifrices.

Anisole (methoxybenzene). A colourless liquid aromatic ether, $C_6H_5OCH_3$, used chiefly as a solvent. M.pt. -37°C; b.pt. 155°C; r.d. 0.99.

Anisotropic. Having different properties in different directions. For example, an anisotropic crystal has different physical properties (refractive index, modulus of elasticity, etc.) along different axes because of its crystal structure. *Compare* isotropic.

Annealing. Process in which glass, metals, and other materials are treated to render them less brittle and more workable. The material is heated and then cooled very slowly and unformly, with the time and temperature set according to the properties desired. Annealing increases ductility and relieves internal strains that lead to failures in service.

Annihilation. Collision of a particle and its antiparticle resulting in conversion of the particles into electromagnetic radiation (*annihilation radiation*). For example, an electron and a positron can collide to yield two gamma-ray photons. The total energy of the electromagnetic radiation is equivalent to the total mass of the particles according to the law of conservation of mass-energy.

Annual. Plant that propagates itself by seed and undergoes its entire life cycle within one growing season, as distinguished from a Biennial or Perennial. Some cultivated annuals will blossom during a season only if started under glass and set out as young plants; others bloom where sown.

Annual rings. Growth layers of Wood produced yearly in the stems and roots of trees and shrubs. When well-marked alteration of seasons occurs (e.g., either cold and warm or wet and dry), a sharp contrast exists between the early and late-season growth—the wood cells are larger earlier when growing conditions are better. In uniform climates, there is little visible difference between annual rings. The number of annual rings reflects the age of a tree; the thickness of each ring reflects environmental and climatic conditions.

Annulus. 1. A two-dimensional hollow ring; the plane figure bounded by two concentric circles. 2. A solid equivalent to a solid cylinder with a hole along its axis.

Anode. An electrode with a positive electric charge, as in a cell, thermionic valve, rectifier, etc. *Compare* cathode.

Anodizing. The process of forming an oxide layer on aluminium by making it the anode in an electrolytic bath containing sulphuric or chromic acid solution. The layer of aluminium oxide is porous and can be coloured by certain dyes.

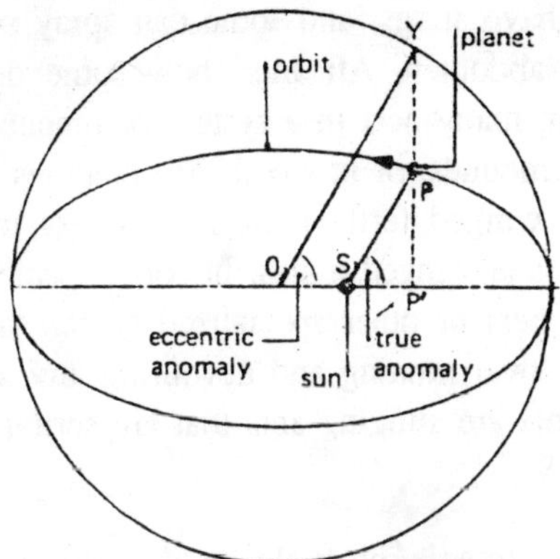

Fig. A-12. Anomaly.

Anomaly. An angular measure used in determining the position of a planet in its orbital path about the sun. *True anomaly* is the angle between the planet's perihelion, the sun, and the planet itself measured in the direction of the planet's motion. *Mean* anomaly is the angle between the perihelion, the sun, and an imaginary planet having the same period as the planet in question but moving at a constant velocity. (see Kepler's Law) A third type, *eccentric anomaly,* is based on the planetary orbit conceived as a circle having the major axis of the planet's true *elliptical* orbit as its diamcter. The eccentric anomaly is the angle P'OY in the illustration.

Anorexia Nervosa. Disorder involving abnormal loss of appetite and refusal to eat, occurring most commonly among adolescent women. Because of complex emotional disturbances, the patient becomes overconcerned with weight and diet, fears fatness, and has a distorted body image. With continuous dieting she becomes emaciated, loses body hair, and ceases to menstruate. Some die of self-starvation, but most either recover completely, particularly after undergoing psychotherapy, or more typically, have repeated relapses and recoveries.

Ant. Insect (family Formicidae) belonging to the same order as the Bee and Wasp. Ants have a narrow stalk joining the abdomen and thorax, and biting mouthparts (sometimes used for defense). Some ants have stings, and some can spray poison from the end of the abdomen. All ants show some degree of social organization; many nest in a system of tunnels in the soil, in constructed mounds, or in wood. Ant colonies usually include three castes: winged, fertile females; wingless, infertile females, or workers; and winged males. In some species workers may become soldiers or other specialized types. Army ants travel in columns, over-running and devouring any animals in their path; fire ants are stinging ants that are serious pests in parts of the S U.S..

Antarctic Circle. Imaginary circle on the earth's surface at 66° 30'S lat., marking the southernmost point at which the sun can be seen at the summer Solstice (about June 22) and the northernmost point of the southern polar regions at which the midnight sun is visible.

Anteater. Toothless Mammal of three genera of the order Edentata found in tropical Central and South America. It feeds on ants, termites, and other insects. The great ant-eater, or ant bear (*Myrmecophaga*), has an elongated snout, a coarse-haired body about 4 ft (1.2 m) long, and a long, broad tail; the arboreal collared, or lesser, anteater (*Tamandua*) is less than half the size of the great anteater; and the arboreal two-toed anteater (*Cyclopes*) is about the size of a squirrel.

Antelope. Any of various hoofed ruminant Mammals of the Cattle family. True antelopes are found only in Africa and Asia. Antelopes usually stand 3-4ft (90-120 cm) at the shoulder, but range from the 12-in. (30-cm) pygmy antelope to the 6-ft (180-cm) giant eland. Antelope horns, unlike Deer horns, are unbranched and are not shed. Species of antelope include the bushbuck, with its spiral horns and oxlike body; the addax, a large desert antelope of N. Africa; the horselike oryx, found in Africa and arabia; the GNU and its close relative the hartebeest, a swift, horselike animal with U-shaped horns; and the Gazelle.

Antenna or **Aerial.** System of wires or other conductors used to transmit or receive radio or other electromagnetic waves. In a transmitting antenna, the signal from an Electric Circuit causes electrons in the antenna to oscillate; these moving electric charges generate Electro Magnetic Radiation, which is then transmitted through space. The distribution pattern of the transmitted wave depends on the design of the antenna; radio broadcast-station antennas are frequently designed to emit waves in all dirctions, whereas those used for Radar and for certain communication systems focus the waves in a single direction. In a receiving antenna, electromagnetic waves cause the electrons in the antenna to oscillate, inducing a signal that can be detected by an electric circuit. In general, a longer antenna is used to transmit or receive signals of longer wave-lengths. Theoretically, the same antenna can be used both for sending and for receiving signals, but in practice, transmitting antennas are constructed to handle higher power loads than receiving antennas.

Anthracene. A colourless crystalline solid tricyclic aromatic hydrocarbon $C_{14}H_{10}$, obtained from coal tar and used as an important source of dyestuffs. M.pt. 218°C; b.pt. 340°C; r.d. 1.25.

Anthrax. Infectious disease of animals that can be transmitted from animals to humans through contact or the inhalation of spores. Primarily affecting sheep, horses, hogs, cattle, and goats, anthrax is usually restricted to people who handle hides of animals (e.g., farmers, butchers, and veterinarians) or sort wool. The disease, usually fatal to animals but not humans, is caused by *Bacillus anthracis,* discovered by Robert Koch in 1876. Louis Pasteur developed a method of vaccinating sheep and cattle against anthrax, and the disease is now relatively rare in the U.S.

Anthropology. Study of the origin, development, and varieties of human beings and their societies. Emerging as an independent science in the late 18th cent., it developed two main divisions; physical anthropology, which focuses on human Evolution and variation, using methods of Physiology, anthropometry, Genetics, and Ecology; and cultural anthropology, which includes Archaeology, ethnology, social anthropology, and Linguistics.

Antianxiety drug. Drug administered for the relief of Anxiety; in larger doses, antianxiety drugs produce sleep, sedation, and anesthesia. Those frequently prescribed in the U.S. include diazepam (Valium), chlordiazepoxide (librium), and meprobamate (Miltown). Side effects include drowsiness and, with prolonged use, possible dependence.

Antibiotic. Any of a variety of substances, usually obtained from microorganisms, that inhibit the growth of or destroy certain other microorganisms. The foundation for the development and understanding of antibiotics was laid during the 19th cent. when Louis Pasteur proved that one species of microorganism can kill another and paul Ehrlich developed the idea of selective toxicity: that a specific substance can be toxic to some organisms, e.g., infectious bacteria, but harmless to others, e.g., human hosts. Further pioneering work in the 20th cent. by Alexander Fleming, Rene Dubos, and Selman Waksman led to the discovery of Penicillin (1939) and streptomycin (1944). Mass production of antibiotic drugs began during World War II with streptomycin and penicillin. Today antibiotics are produced by various methods, including staged fermentation in huge tanks of nutrient media, synthesis in the laboratory, and chemical modification of natural substances. Antibiotics can be classified according to chemical structure, microbial origin, mode of action, or effective range. The tetracyclines are broad-spectrum drugs, effective against a wide range of bacteria (both gram-positive and gram-negative See Groms stain), rickettsias, and the psittacosis Virus. The antibiotics bacitracin, penicillin, the erythromycins, and the cephalo-sporins work primarily against Gram-positive organisms. Polymyxins and other narrow-range antibiotics are effective against only a few species. Antibiotic drugs may be injected, given orally, or applied to the skin. Some, like penicillin, are allergenic, and can cause rashes or shock. Others, like the tetracylines, can alter the intestinal environment, encouraging superinfection by fungi or other microorganisms. Many antibiotics are less effective than formerly because resistant strains of microorganisms have evolved. Antibiotics have been used to enhance the growth of animals used as food, but some authorities question the practice because it encourages the development of resistant strains of bacteria infecting animals, and because continuous low exposure to antibiotics can sensitize human beings, making them unable to take such drugs later to treat infection.

Antichlor. A substance used for removing chlorine from a solution, material, etc.

Anticoagulant. Substance, such as heparin or derivatives of coumarin, that inhibits blood clotting. Anticoagulants are used to treat blood clots in leg and pelvic veins, in order to reduce the risk of the clots traveling and obstructing blood flow to vital organs (e.g., the heart and lungs) and causing Thrombosis or Stroke.

Antidepressant. Drug used to treat psychic depression. Antidepressants, such as imipramine, are given to elevate mood, counter suicidal thoughts, and increase the effectiveness of psychotherapy, although the exact mechanisms of their action are unknown. Side effects include dry mouth, dizziness, and fatigue.

Antiderivative. The function whose derivative is the given function. Thus, the antiderivative of 2x is x^2 since the derivative of x^2 is 2x.

Antiferromagnetism. A type of magnetic behaviour in which the material has a very low susceptibility that increases with temperature. It is found in certain inorganic solids, such as MnO and FeO, having lattices in which adjacent atoms have antiparallel spins. Above a certain temperature, the *Neel temperature,* the susceptibility falls as the material becomes paramagnetic.

Antifreeze. Substance added to a solvent to lower its freezing point. Antifreeze is typically added to water in the cooling system of an internal combustion engine so that it may be cooled below the freezing point of pure water (32° F or O°C) without freezing. Automotive antifreezes include ethylene glycol (the most widely used), methanol, ethanol, isopropyl alcohol, and propylene glycol.

Antilogarithm. The number whose logarithm is the given number. Thus, if $x = 10^y$, so that $y = \log x$ to base 10, then x is the antilogarithm (written antilog, or $\log^{-1}$) of y to base 10.

Antimatter. Material composed of antiparticles, which correspond to ordinary protons, electrons, and neutrons but have the opposite electrical charge and magnetic moment. When matter and antimatter collide, both may be annihilated, and other Elementary Particles, such as photons and pions, are produced. In 1932 Carl D. Anderson, while studying cosmic rays, discovered the positron, or antielectron, the first known antiparticle. Any antimatter in our part of the universe is necessarily very short-lived because of the over-whelming preponderance of ordinary matter, by which the antimatter is quickly annihilated.

Antimony. Symbol: Sb. A brittle bluish-white metalloid element occurring in several minerals, especially stibnite, Sb_2S_3. It is extracted by roasting to produce the oxide, which is reduced with carbon. Besides metallic antimony there are three other nonmetallic allotropes. The element is used for increasing the hardness of lead that is used in accumulators, type metal, and sheathing cables. The chemistry of antimony is similar to that of arsenic. It forms covalent antimony III (*antimonous*) and antimony V (*antimonic*) compounds. Ionic antimony III compounds, containing the Sb^{3+} ion, also exist. In addition the element has *antimonyl* compounds, containing the SbO^{+} ion. A.N. 51; A.W. 121.75; m.pt. 650°C; b.pt. 1380°C: r.d. 6.69.

Antimony Oxide. Either of two amphoteric oxides of antimony. *Antimony (III)* oxide (antimony trioxide) is a white solid, Sb_2O_3, obtained by roasting antimony ores and used in flameproofing textiles, as a mordant, and as a white pigment (*antimony white*). M.pt. 450°C; r.d. 5.6 *Antimony (V)* oxide (antimony pentoxide) is a yellow solid. Sb_2O_5, made by the action of nitric acid on the metal. It decomposes to the III oxide on heating. R.d. 5.6.

Antinode. A point of maximum displacement in a standing wave.

Antiparallel. Designating vectors that have parallel lines of action and act in opposite directions.

Antiparticle. An elementary particle that has the same mass as a given particle and a charge, baryon number, and isospin quantum number of the same magnitude but of opposite sign. Thus, the positron is the antiparticle of the electron; the antiproton, with negative charge, is the antiparticle of the proton. Collision between a particle and its antiparticle can result in annihilation.

Anxiety. Anticipatory tension or vague dread persisting in the absence of a specific threat. It is characterized by increased pulse rate and blood pressure, quickened breathing, perspiration, and dryness of the mouth. It has been defined as the tension resulting from a sense of failure in interpersonal relations or as a reaction to unresolved conflicts that threaten the personality.

Apatite. Phosphate mineral [(Ca, Pb)$_5$ (PO$_4$)$_3$ (F, Cl, OH)], transparent to opaque in shades of green, brown, yellow, white, red, and purple. Apatite, a minor constituent of many types of rock, is mined in Florida, Tennessee, Montana, N Africa, Europe, and the USSR. Large deposits are mined for use in making phosphatic fertilizers, and two varieties are used to a limited extent in jewelry.

Ape. Primate of the family Pongidae, or Simiidae, closely related to human beings. The small apes, the Gibbon and siamang, and the smallest of the great apes, the Orangutan, are found in Se Asia; the other great apes, the Gorilla and the Chimpanzee, are found in Africa. They vary in size from the 3-ft (1.8-m), 15-lb (6.8-kg) gibbon to the 6-ft (1.8-m), 500-lb (227-kg) gorilla. All are forest dwellers and spend at least some time in trees; unlike Monkeys, they can swing hand-over-hand. Their brains, similar in structure to the human brain, are capable to fairly advanced reasoning.

Aperture. The part of a mirror or lens that reflects or admits light in an optical system. The aperture of a lens or curved mirror is measured by the diameter of the refracting or reflecting surface.

Aperture Synthesis. A technique used in radio astronomy to obtain observations on a large area of the sky using an array of small radio telescopes. The array of antennas is moved into a large number of different positions and observations made in each position are recorded. The data can be used to construct a map of radio activity over a wide area of the sky—the result is equivalent to using a single radio telescope with a very large aperture.

Aphasia. Language disturbance caused by a lesion of the brain, causing partial or total impairment in the individual's ability to speak, write, or comprehend the meaning of spoken or written words. Treatment consists of reeducation; the methods used include those employed in the education of the deaf.

Aphelion. The point of furthest recession from the sun attained by a planet, asteroid, comet, or other orbiting body. *Compare* perihelion.

Aphid or **Plant Louse.** Tiny, usually green, pear-shaped Insect injurious to vegetation. Aphids feed by sucking plant juices through their beaks. Most species live in large colonies; some species live symbiotically with ants. Many inhabit swellings of plant tissues (galls) that they form with their secretions.

Aplanatic. Denoting a lens or other optical system that is free from spherical aberration.

Apogee. The point at which the moon or any other orbiting satellite is furthest away from the earth. *Compare* perigee.

Apothecaries weight. A system of weights based on the grain.

Appendix. Small, worm-shaped blind tube projecting from the large intestine into the lower right abdominal cavity. It has no function and is considered a remnant of a previous digestive organ. Appendicitis can occur if accumulated and hardened waste matter becomes infected; rupture of such an appendix can spread infection to the peritoneum (abdominal membrane), causing peritonitis.

Apple. Any tree (and its fruit) of the genus *Malus* of the Rose family. The common apple (*M. Sylvestris*) is the best-known and commercially most important temperate fruit. It is native to W. Asia, but has been widely cultivated from prehistoric times. Thousands of varieties exist, e.g., Golden Delicious, Winesap, Jonathan, and McIntosh. The fruit is consumed fresh or cooked, or is used for juice. Partial fermentation of apple juice (sweet cider) produces hard cider (from which applejack liquor is made); fully fermented juice yields vinegar. The hardwood is used in cabinetmaking and as fuel. The fruit of crab apple trees (which are cultivated as ornamentals) is used for preserves and jellies.

Appleton Layer. See ionosphere. [After Sir Edward Victor Appleton (1892-1965), English physicist].

Apricot. Tree (*Prunus armeniaca*) of Rose family and its fruit, native to Asia. In the U.S., it cultivated chiefly in California. the fruit is used raw, canned, preserved, and dried, and in making a cordial and a brandy.

Apsis. Point in the Orbit of a smaller body where it is at its greatest or least distance from a larger body to which it is attracted. In an elliptical orbit these points are called the apocenter and pericenter; corresponding terms for elliptical orbits around the sun, the earth, and a star are, respectively, aphelion and perihelion, apogee and perigee, and apastron and periastron. The line of apsides, an imaginary straight line connecting the two apsides of an elliptical orbit, may shift because of gravitational influences of other bodies or relativistic effects (See Relativity).

Aquadag. R A colloidal suspension of graphite in water, used as a lubricant and for producing conductive graphite coatings.

Aquamarine. Transparent blue to bluish-green variety of the mineral Beryl, used as a Gem. Sources include Brazil, the USSR, Madagascar, and parts of the U.S. Oriental aquamarine is a transparent bluish variety of Corundum.

Aqua Regia. A fuming yellow corrosive mixture of one part of nitric acid to three or four parts hydrochloric acid. It is used in metallurgy, since it dissolves all metals, even gold. The mixture contains chlorine and nitrosyl chloride (NOCl).

Aquarium. Name for any supervised exhibit of living aquatic animals and plants. Aquariums are known to have been built in ancient Rome, Egypt, and the Orient. large, modern public aquariums were made possible by the development of glass exhibit tanks capable of holding over 100,000 gal (378,500 liters) of water. Aquarium maintenance requires careful regulation of temperature, light, food, oxygen, and water flow; removal of injurious waste and debris; and attention to the special requirements of the individual species. Freshwater and saltwater aquariums are often maintained for research and breeding purposes by universities, marine stations, and wildlife commissions.

Aquatint. Etching technique. A metal plate is coated with resin through which acid bites an evenly pocked surface. When printed, it produces tonal effects that resemble wash drawings. Said to have been invented in the 1760s by J.B. Le Prince, aquatint is often combined with other types of etchings, as in Goya's series of mixed aquatint etchings.

Aqueduct. Conduit for conveying water, built chiefly to bring fresh drinking water into cities. Aqueduct types include tunnels cut through rock, pipelines made from various materials, and the conduits that top the arched, bridgelike structures that stand in many parts of Europe and are legacies of the ancient Roman water supply system. Modern aqueducts utilize gravity, pumps, and the propulsive force of very high pressures in underground tunnels to move water over great distances.

Aqueous. Denoting a solution in which water is the solvent.

Aqueous Humour. *See* Eye.

Arachnid. Mainly terrestrial Arthropod of the class Arachnida, including the Spider, Scorpion, Mite, tick, and Daddy Longlegs. The arachnid's body is divided into a cephalothorax with six pairs of appendages and an abdomen. Most are carnivorous, and some use the first two pairs of appendages to crush and kill prey; the others are for walking. Arachnids have simple eyes and sensory bristles; they have no antennae. Respiration is through air tubes or primitive structures called book lungs.

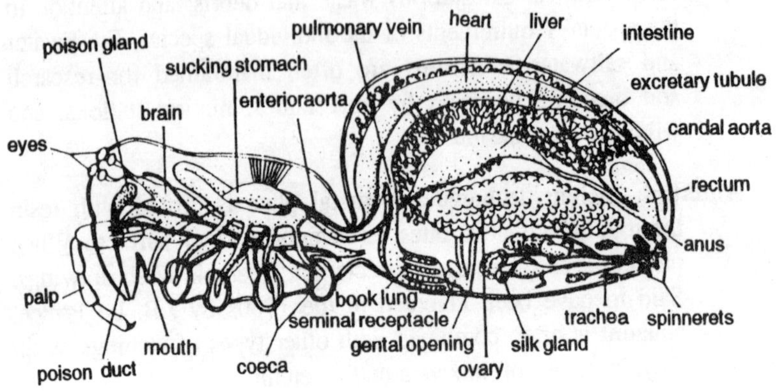

Fig. A-13. Arachnid. Internal anatomy of a spider, a representative arachnid.

Arborvitae. Aromatic evergreen tree (genus *Thuja*) of the cypress family, native to Asia and North America. Arborvitaes have tiny cones and flattened branchlets with scalelike leaves. The white cedar (*T. occidentalis*), popular for hedges, and the red cedar (*T. plicata*), important for lumber, both have decay-resistant wood.

Arc. 1. An open segment of a curve.

2. A luminous electrical discharge between two electrodes, characterized by a high current and a low potential difference. The electrodes are heated by the discharge and their evaporation helps maintain it.

3. Denoting the inverse of a circular or hyperbolic function. Thus the arc tangent of a number is any angle whose tangent is the given number, denoted arc tan x or $\tan^{-1} x$.

Archaeology. Scientific study of material remains of human cultures to derive knowledge about prehistoric times and to supplement documentary evidence of historic times. Research into the life and culture of the past began in 15th-cent. Italy with the excavation of ancient Greek sculpture. It was advanced in the 18th cent. by excavations at Herculaneum and Pimpeii. In the 19th cent. public interest was stimulated by the acquisition of the Elgin Marbles and by the excavations of Heinrich Schliemann at Troy and in Greece and those of Arthur Evans at Crete; Egyptology was born with the recovery of the Rosetta stone. Stratigraphic excavation of such sites as Barrows, mounds, and kitchen middens unearthed levels of material culture and of the past. Scientific analysis of these material remains led Danish archaeologist Christian Thomsen to classify cultures according to the principal materials used for weapons and tools; Stone Age, Bronze Age, Iron Age. Realizing that an exact equation of technical tradition with cultural development was too rigid, later interpreters emphasized Ecology and food production. Improved field techniques, new Dating methods, and the study of existing aboriginal groups have enhanced the knowledge of prehistory. Important archaeological work is currently being done in Africa, (see Leakey family), China, and the Americas.

Archimedes (arkime dez), C 287 B.C.-212 B.C., Greek mathematician, physicist, and inventor. His reputation in antiquity was based on several mechanical contrivances, e.g., Archimedes' Screw that he is alleged to have invented. One legend states that during the Second Punic War he protected his native Syracuse from the besieging armies of Marcus Cludius Marcellus for three years by inventing machines of war, e.g., various ballistic instruments and mirrors that set Roman ships on fire by focusing the sun's rays on them. In modern times, however, he is best known for his work in mathematics, mechanics, and hydrostatics. In mathematics, he calculated that the value of Π is between $3^{10/71}$ and $3^{1/7.}$ devised a mathematical exponential system to express extremely large numbers; proved that the volume of a sphere is two thirds the volume of a circumscribed cylinder; and, in calculating the areas and volumes of various geometrical figures, carried the method of exhaustion invented by Eudoxus of Cnidus far enough in some cases to anticipate the invention (17th cent.) of the Calculus. One of the first to apply geometry to mechanics and hydrostatics, he proved the law of the lever entirely by geometry and established Archimedes' Principle. In another legendary story, the ruler Hiero II requested him to find a method for determining whether a crown was pure gold or alloyed with silver. Archimedes realized, as he stepped into a bath, that a given weight of gold would displace less water than an equal weight of silver (which is less dense than gold); and he is said, in his excitement at his discovery, to have run home naked, shouting *"Eureka! Eureka!"* ("I have found it! I have found it!"). He was killed by a Roman Soldier, supposedly while absorbed in mathematics.

Archimedes' Principle. Principle that states that a body immersed in a fluid is buoyed up by a force equal to the weight of the displaced fluid. The principle applies to both floating and submerged bodies, and to all fluids. It explains not only the buoyancy of ships but also the rise of a helium-filled balloon and the apparent loss of weight of objects underwater.

Archimedes' screw. A simple mechanical device used to lift water and such light materials as grain or sand, and believed to have been invented by Archimedes in the 3rd cent. B.C. It consists of a large continuous screw inside a cylindrical chamber. To lift water—for example, from a river to its bank—the lower end is placed in the river, and water rises up the spiral threads of the screw as it is revolved.

Arctic. The northernmost area of the earth, centered on the North Pole. It can be defined as embracing all lands located N of the Arctic Circle (lat. 66°31'N) or all lands located N of the 50°F (10°C) July isotherm, which is roughly equivalent to the tree line. It therefore generally includes the Arctic Ocean; the northern reaches of Canada, Alaska, the USSR, and Norway; and most of Greenland, Iceland, and Svalbard. Ice sheets and permanent snow cover regions where average monthly temperatures remain below 32°F (0°C) all year; Tundra, which flourishes during the short summer season, covers areas where temperatures are between 32°F and 50°F (0°—10°C) for at least one month. Daylight ranges from 24 hours of constant daylight at the Arctic Circle to six months at the North pole in summer; conversely, in winter, nights last from 24 hours of darkness at the Arctic Circle to six months at the North Pole. The Arctic is of great strategic value as the shortest route between the U.S. and USSR. Its rich oil and natural gas deposits have been discovered since the International Geophysical Year (1957-58) on Alaska's North Slope, Canada's Ellesmere Island (1972), and the northern areas of Siberia, in the USSR. The first explorer to reach the North Pole was Robert E. Peary, in 1909.

Arctic Circle. Imaginary circle on the earth's surface at 66° 31'N lat., marking the northernmost point at which the sun can be seen at the winter Solstice (about Dec. 22) and the southernmost point of the northern polar regions at which the midnight sun is visible.

Are. Symbol: a. A metric unit of area equal to 100 sq. metres. it is equivalent to 119.60 sq. yards.

Area. A measure of a two-dimensional surface. Units of area are the square of any unit length, e.g. square centimetres (cm^2) or square feet (ft^2).

Arecibo Ionospheric Observatory. Radio-astronomy facility (completed 1963) located at Arecibo, Puerto Rico; it is operated by Cornell Univ. under contract with the U.S. National Science Foundation. Its fixed antenna of spherical section, 1,000 ft. (305 m) in diameter, is the largest radio-telescope antenna in the world.

Argand Diagram. A representation of a complex number as a point on a plane (*see* diagram). With Cartesian coordinates, the complex number x + iy is represented by the point (x,y).

In polar coordinates, this point is (r,θ) where θ is the argument of the complex number and r the modulus. [After Jean Robert Argand (1768-1822), French mathematician born in Geneva.]

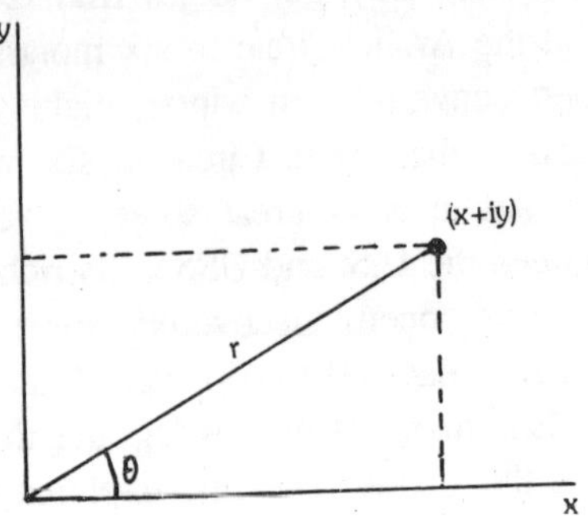

Fig. A-14. Argand diagram.

Argentic. *See* Silver.

Argentite (silver glance). A grey-black mineral consisting of silver sulphide, Ag_2S, used as an ore of silver.

Argentous. *See* Silver.

Argol (tartar). A crude form of potassium hydrogen tartrate deposited as a reddish-brown solid on the side of wine vats.

Argon. Symbol : A. A colourless odourless tasteless nonflammable gaseous element present in the atmosphere (0.94%) from which it is obtained as a by-product in the liquefaction of air. Argon is used for filling fluorescent lamps and electric-light bulbs, and in argon-arc welding. It is one of the inert gases. A.N. 18; A.W. 39.948; m.pt.-189.3°C; b.pt. -185.8°C; r.d. 1.38 (air = 1).

Argument. 1. (of a function). A value of an independent variable variable yielding some particular value of the function. The function $y = x^2$ takes the value 4 in the field of rational numbers for arguments +2 and -2.

2. (of a complex number). The angle between the line joining the representation of a complex number on the Argand diagram with its origin, and the x-axis, normally measured in radians. The complex number can also be represented uniquely in terms of its argument and its modulus. If $x + iy$ has modulus θ, then $\tan \theta = y/x$.

Aristotle. 384-322 B.C., Greek philosopher. He studied (367-347 B.C.) under Plato and later (342-339 B.C.) tutored Alexander The Great at the Macedonian court. In 335 B.C. he opened a school in the Athenian Lyceum. During the anti-Macedonian agitation after Alexander's death Aristotle fled (323 B.C.) to Chalcis, where he died. His extant writings, largely in the form of lecture notes made by his students, include the *Organum* (treatises on logic); *Physics; Metaphysics; De Anima* [on the soul]; Nicomachean Ethics and *Eudemian Ethics; Politics; De Poetica; Rhetoric;* and works on biology and physics. Aristotle held philosophy to be the discerning through the use of systematic Logic as expressed in Syllogisms, of the self-evident, changeless first principles that form the basis of all knowledge. He taught that knowledge of a thing requires an inquiry into causality and that the "final cuase"—the purpose or function of the thing—is primary. The highest good for the individual is the complete exercise of the specifically human function of rationality. In contrast to the Platonic belief that a concrete reality partakes of a form but does not embody it, the Aristotelian system holds that, with the exception of the Prime-Mover (God), form has no separate existence but is immanent in matter. Aristotle's work was lost following the decline of rome but was reintroduced to the West through the work of Arab and Jewish scholars, becoming the basis of medieval Scholasticism.

Arithmetic. Branch of mathematics (and the part of Algebra) concerned with the fundamental operations of addition, subtraction, multiplication, and division of numbers. Conventionally, the term *arithmetic* covers simple numerical skills used for practical purposes, e.g., computation of areas or costs. The study of arithmetic also deals abstractly with the laws and properties (e.g., the Associative Law, the Commutative Law, and the Distributive Law) governing the various operations.

Arithmetic. The elementary study of the properties of and relations between the integers or the rational numbers under the operations of addition, substraction, multiplication, and division.

Arithmetic mean. The result of dividing the sum of a given set of numbers by the number of numbers in the set.

Arithmetic progression. A sequence in which each term differs from the preceding one by a fixed number. An arithmetic progression has the form $a, (a + d), (a + 2d) \ldots [a + (n - 1)d]$, where d is the common difference and n is the number of terms. The *arithmetic* series formed from this sequence has a sum equal to $(n/2[2a + (n - 1)d]$. *Compare* geometric progression.

Armature. 1. The part of a generator in which the electric current is induced or the corresponding part of an electric motor in which the driving current flows.

2. Any of various parts of electrical equipment in which an electric voltage is induced, as in a loudspeaker or record-player pick-up.

3. A piece of iron or steel held across the poles of an electromagnet so that it can be used to apply a force.

Aromatic. Denoting a chemical compound that has the property of aromaticity, as characterized by benzene. The term was originally used in a more literal sense to distinguish fragrant compounds from aliphatic ("fatty" compounds).

Aromatic compound. Any of a large class of organic compounds including Benzene and compounds that resemble benzene in chemical properties. Aromatic compounds contain usually stable ring structures, often made up of six carbon atoms arranged hexagonally. Some of the compounds, however, have rings with more or fewer atoms, not necessarily all carbon. Furan, for example, has a ring with four atoms of carbon and one of oxygen. Also, two or more rings can be fused, as in naphthalene. The characteristic properties of the class, notably the stability of the compounds, derive from the fact that aromatic rings permit the sharing of some electrons by all the atoms of the ring which increases the strength of the bonds.

Aromaticity. The property of certain chemical compounds of having unsaturated molecules yet undergoing substitution reactions rather than addition reactions. Such compounds are said to be *aromatic*. The archetypal example is benzene, which, in the nineteenth century, was the subject of considerable speculation. The basic problem— the *benzene problem*— was that of reconciling the formula of benzene, C_6H_6, with its chemical reactions. The empirical formula is the same as that of acetylene, C_2H_2, and it might be expected that benzene would undergo similar reactions. However, benzene does not show the usual behaviour of a compound containing double or triple bonds.

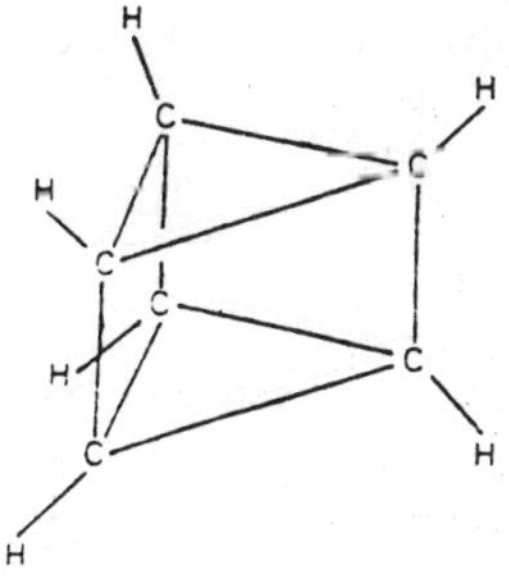

Fig. A-15. Ladenburg benzene.

For example, acetylene adds bromine to yield CHBr: CHBr and eventually $CHBr_2$ $CHBr_2$: benzene, with an iron III bromide catalyst suffers displacement of one of its hydrogen atoms to yield C_6H_5Br. This type of activity indicates that benzene might be saturated and a postulate structure is that of *Ladenburg benzene.*

Benzene, however, does not always act as a saturated compound. In sunlight bromine is added to give $C_6H_6Cl_6$ and hydrogen can also be added with a nickel catalyst to yield cyclohexane, C_6H_{12}. Some of the postulated structures accounting for this behaviour are shown in the first illustration.

In 1865 the German chemist August Kekule (1829-96) suggested a structure with alternate double and single bonds in a hexagonal ring. To account for the fact that benzene has only three disubstitution products, he further proposed that the positions of the bonds oscillate, so that two molecules are in equilibrium. This structure—the *Kekule formula*—is the one usually used in formulas of compounds containing benzene rings.

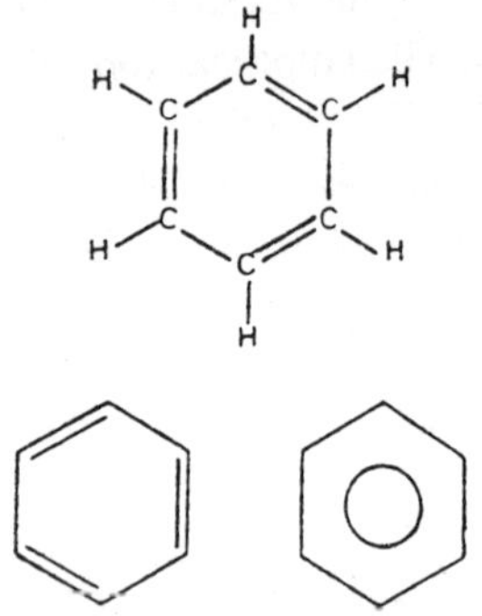

Fig. A-16. Kekule formula of benzene .

The modern idea of aromaticity is based not on equilibrium between Kekule structures but on resonance between them. The bonds in benzene have characters between double and single bonds; the carbon atoms are held together by six single bonds and the remaining six electrons, from the double bonds, are delocalized over the ring. This is the reason benzene has all its C-C bonds of the same length and undergoes electrophilic substitution reactions. The German chemist Erich Huckel (b. 1896) suggested that for a compound to be aromatic it had to have a planar ring of alternating double and single bonds with $4n + 2$ pi electrons available from the double bonds for delocalization. This is the *Huckel rule.* In benzene, $n = 1$: i.e. there are 6 pi electrons delocalized. The Huckel rule is often used as the definition of aromaticity. *See also* pseudoaromatic.

Aromatization. The conversion of an aliphatic ring compound into an aromatic compound by dehydrogenation. A typical method is catalytic dehydrogenation using a catalyst such as nickel, platinum, or palladium. Cyclohexane, for example, will yield benzene: $C_6H_{12} = C_6H_6 + 3H_2$.

Arrhenius equation. The equation $V = A \exp (E_A/RT)$, relating the rate constant v of a chemical reaction to the activation energy E_A . A is a constant for a given reaction. [After Svante Arrhenius (1859-1927), Swedish chemist].

Arrhenius, Svante August (Ara' neas), ·1859-1927), Swedish chemist. For his theory of electrolytic dissociation, or ionization he won the 1903 Nobel Prize in chemistry. He also investigated osmosis and toxins and antitoxins. He became (1905) director of the Nobel Institute for Physical Chemistry, Stockholm.

Arrowroot. Plant (genus *Maranta)* of the family Marantaceae, found mainly in warm swampy forests of the Americas. The term *arrowroot* also applies to the easily digested starch obtained from the true, or West Indian, arrowroot *(M. arundinacea)*. Plants from other families produce similar starches, e.g., East Indian arrowroot (Ginger family); Queensland arrowroot (Canna family); Brazilian arrowroot, or tapioca (Spurge family; Florida arrowroot, or sago.

Arrowworm. Common name for Chaetognatha, a phylum of slender-bodied, transparent marine invertebrate animals widely distributed, but preferring warm, shallow seas. Most are less than 1 in. (2.5 cm) long. The head is well-developed, with eyes and other sensory organs, grasping spines, teeth, and a protective hood. Arrowworms reproduce sexually.

Arsenate. Any salt of arsenic acid.

Arsenic. Symbol: As. A silvery-grey brittle metalloid element occurring in several minerals, including orpiment, realgar, and arsenopyrite. It is also found in copper and lead ores, the main source being arsenic oxide recovered as a by-product in the smelting process. This is reduced with charcoal to give the metallic element. Two other allotropic forms exist: yellow arsenic and black arsenic. Metallic arsenic is used in some alloys and semiconductors. It is a reactive element, combining directly with the halogens and tarnishing in air. Like phosphorus it exhibits a valency of three *(arsenious or arsenous)* and five *(arsenic)*. The term "arsenic" is sometimes loosely applied to the oxide, As_2O_3. A.N. 33; A.W. 74.9216; m.pt. 218°C (28 atm); sublimes at 615°C; r.d. 5.72 (metallic).

Arsenic acid. A white solid tribasic acid, H_3AsO_4. $1/2H_2O$, made by the action of hot concentrated nitric acid on arsenic III oxide.

Arsenic Oxide. Either of two oxides of arsenic. *Arsenic (III) oxide* (arsenic trioxide) is a white odourless tasteless poisonous powder, As_2O_3, manufactured by roasting arsenopyrite ore. It ia an amphoteric compound, the anhydride of arsenious acid, and is used in the preparation of other arsenic compounds. It is also used in insecticides, weedkillers, rat poison, and the manufacture of glass. Arsenic (III) oxide is often loosely called *arsenic*. Sublimes at 193°C; r.d. 3.87. *Arsenic (V) oxide* (arsenic pentoxide) is a white deliquescent amorphous solid, As_2O_5, prepared by oxidizing the lower oxside. It is used in dyeing. Printing, and insecticides and is the anhydride of arsenic acid. M.pt. 315°C; r.d. 4.09.

Arsenious. *See* arsenic.

Arsenious acid. An acid, H_3AsO_3, obtained in aqueous solutions of arsenic III oxide.

Arsenite. Any salt of arsenious acid.

Arsenopyrite (mispickel). A white or grey mineral consisting of a mixed sulphide and arsenide of iron, FeS_2. - $FeAs_2$. It is an ore of arsenic and also contains small amounts of gold.

Arsenous. *See* arsenic.

Arsine. A colourless poisonous gas, AsH_3, with an odour of garlic. It is an unstable compound, decomposing readily to arsenic and hydrogen. M.pt. - 113.5°C; b.pt, -55°C.

Arteriosclerosis. General term for a condition characterized by thickening, hardening, and loss of elasticity of the walls of the arteries. In its most common form, atherosclerosis, fatty deposits, e.g., Cholesterol, build up on the inner artery walls; in some cases calcium deposits also form. The blood vessels narrow, and blood flow decreases; Thrombosis, Heart Disease, and Stroke may result. Surgical treatment is sometimes effective, but there is no specific cure. A low-cholesterol diet and control of predisposing factors, such as Hypertension, smoking, Diabetes, and obesity, are usually recommended.

Artesian well. Deep well drilled into an inclined aquifer (layer of water-bearing porous rock or sediment), where water is trapped under pressure between impervious rock layers. When a well is drilled into the aquifer through the overlying impervious rock, pressure forces water to rise in the well. Artesian water is usually desirable for drinking. In North America, artesian systems supply water to the Great Plains and many East Coast cities. The largest artesian system is in Australia.

Arthritis. Inflammation of one or more joints of the body, usually producing pain, redness, and stiffness. It disables more people than any other chronic disorder. A common form is osteoarthritis, a degenerative disease of the joints that commonly occurs with aging. Rheumatoid arthritis, an Autoimmune disease of unknown cause, is a progressive, crippling joint disorder most common in women between 25 and 50. Initial treatment for arthritis includes use of heat, physical therapy, and Aspirin; in more severe cases, gold salts and Cortisone are given, but they often have undesirable side effects.

Arthropod. Invertebrate animal, having a segmented body covered by a jointed exoskeleton with paired jointed appendages; member of the phylum Arthropoda, the largest and most diverse invertebrate phylum. The exoskeleton is periodically shed (molted) to permit growth and Metamorphosis. Arthropods make up more than 80% (800,000) of all known animal species; they include Trilobites, Horseshoe Crabs, Arachnids.

Artichoke. Name of two plants of the Composite family, both having edible parts. The French, or globe, artichoke *(Cynara scolymus),* of Europe, is a thistlelike plant whose immature, globular flower heads are used as vegetables; only the lower parts of the fleshy bracts ("leaves") and the center ("heart") are eaten. The other artichoke plant is the Jerusalem Artichoke.

Artificial insemination. Technique of artificially injecting sperm-containing semen from a male into a female to cause pregnancy. The technique is widely used in the propagation of cattle, especially to produce many offspring from one prize bull. Artificial insemination is also used in humans when normal fertilization cannot be achieved, as with Sterility or Impotence in the male or anatomical disorders in the female.

Artificial Intelligence. (AI), The use of Computers to model the behavioural aspects of human reasoning and learning. Research in AI is concentrated in some half-dozen areas. In *problem solving,* one must proceed from a beginning (the initial state) to the end (the goal state) via a limited number of steps; AI here involves an attempt to model the reasoning process in solving a problem, such as the proof of a theorem in Euclidean Geometry. In *game theory,* the computer must choose among a number of possible "next" moves to select the one that optimizes its probability of winning; this type of choice is analogous to that of a chess player selecting the next move in response to an opponent's move. In *pattern recognition,* shapes, forms, or configurations of data must be identified and isolated from a larger group; the process here is similar to that used by a doctor in classifying medical problems on the basis of symptoms. *Natural language processing* is an analysis of current or colloquial language usage without the sometimes misleading effect of formal grammars; it is an attempt to model the learning process of a translator faced with the phrase "throw mama from the train a kiss." Cybernetics is the analysis of the communication and control processes of biological organisms and their relationship to mechanical and electrical systems; this study could ultimately lead to the development of "thinking" robots (see Robotics). *Machine learning* occurs when a computer improves its performance of a task on the basis of its programmed application of AI principles to its past performance of that task.

Artificial Respiration. Any measure that causes air to flow in and out of a person's lungs when natural breathing is inadequate or ceases. Respiration can be taken over by mechanical devices such as the artificial lung, in emergency situations, mouth-to-mouth (or, in the case of a small child, mouth-to-nose) procedures are often used. The victim's mouth is cleared of foreign material, and the head is tilted back. Holding the nostrils tightly shut, the person administering artificial respiration places his (or her) mouth over that of the victim and blows air into the lungs, allowing for exhalation after each breath. Twelve vigorous breaths are administered per minute for an adult, twenty shallow breaths per minute for a child.

Artillery. A term now applied to heavy firearms, as distinguished from Small Arms. It came into use in the mid-14th cent. with the introduction in Europe of Gunpowder, which had been discovered many centuries earlier in China. First employed mainly against fortifications, artillery was increasingly used in the field from the early 17th cent. It was characteristically smooth-bore and muzzle-loaded, firing solid, round shot, until the late 19th cent., when breach-loaded, rifled, and shell-firing artillery became standard. Modern artillery includes a variety of long-range guns that fire their shells with rapid muzzle-velocity in a low arc; *howitzers,* which fire on a high trajectory at relatively nearby targets; *antiaircraft guns,* which fire rapidly and at high angles; armor-piercing *antitank guns;* and many field-artillery pieces used in support of infantry and other ground operations. Mobility has become a key factor in the usefulness of heavy firearms, most of which now either are self-propelled or can be towed.

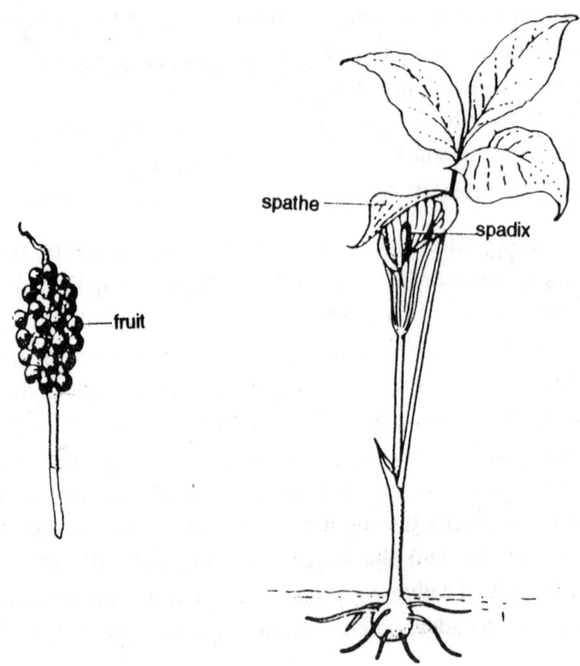

Fig. A-17. Arum: Jack-in-the Pupit, Arisaema triphyllum, a plant of the arum family.

Arum. Common name for the Araceae, a family of chiefly tropical and subtropical herbaceous herbs. The characteristic inflorescence consists of a single fleshy spike (spadix), which bears small flowers, and a typically showy flowerlike bract or modified leaf (spathe) surrounding the spadix. The largest plant inflorescence known belongs to the Sumatran krubi *(Amorphophallus titanum)* of this family; its spadix reaches a height of 15 ft (4.6 m). Among other members of the family are the decorative arum lily, or calla (genus *Zantedeschia);* the smaller, showy water arum, or wild calla (*Calla palustric*); *the climbing shrub philodendron (genus Philodendron*), a popular houseplant; and the decorative *Anthurium* and *Caladium.* Some species in this family have large, starchy edible rootstocks (corms). A major food source in the Pacific and Far East, the corms of elephant's ear *(Colocasia esculenta),* or taro, are the main ingredient of poi. Corms of the arum Indian bread were eaten by natives of E North America.

Aryl. An organometallic compound in which a metal atom is directly bound to an aryl group.

Arylation. The introduction of an aryl group into an organic compound in a chemical reaction.

Aryl Group. An organic group derived by removing a hydrogen atom from an aromatic compound, as in the phenyl group, C_6H_5-.

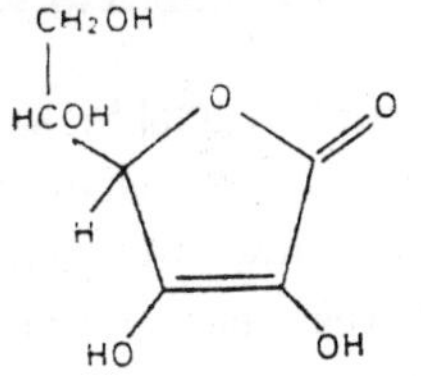

Fig. A-18. Ascorbic acid.

Asbestos. Common name for any of a group of fibrous silicate minerals resistant to acid and fire. Asbestos usually occurs as veins in rocks and seems to be a product of Metamorphism. Chrysotile asbestos ($H_4Mg_3Si_2O_9$), a form of Serpentine, is the main commercial asbestos. Asbestos is produced chiefly in Canada; asbestos products include yarn and rope, pipe covering, brake linings, fire-fighting equipment, and insulating materials. Studies have shown that asbestos particles in the air can cause lung cancer and the lung disease asbestosis.

Ascorbic acid (vitamin C). A white crystalline vitamin, $C_6H_8O_6$, present in many fruits and vegetables. It is also synthesized commercially from D-glucose. Vitamin C is an essential component of the diet of men for the prevention of scurvy. M.pt. 192°C; r.d. 1.65.

Asdic. *See* sonar.

Ash. Tree or shrub (genes *Fraxinus)* of the Olive family, mainly of north temperate regions. The ashes have small clusters of greenish flowers and long-winged, wind-dispersed fruits. The white ash *(F. americana)* is a source of durable hard-wood timber used for sporting goods, furniture, and tool handles; the blue ash *(F. quadrangulata)* provides a blue dye. The mountain ash and prickly ash are not true ashes.

Asimov, Isaac. 1920-, American biochemist and author; b. USSR. He taught biochemistry at Boston Univ., but he is most widely known for his science fiction *(e.g., Robot,* 1950, a short-story collection; and *The Foundation Trilogy,* 1951-53; sequel, 1982 and for popular introductions to science and other topics.

Asparagus. Perennial garden vegetable *Asparagus officinalis)* of the Lily family, native to the E Mediterranean area, cultivated from antiquity and now grown in much of the world. Its green stems function as leaves, and the true leaves are reduced to scales. Its edible shoots are cut in the spring. Related species, the asparagus fern *(A. plumosus)* and smilax *(A. asparagoides),* are used for decoration.

Aspartic Acid. A colourless crystalline optically active nonessential amino acid, $HOOCCH_2CH(NH_2)$ COOH, occurring naturally in sugar cane and sugar beet. It is used in making detergents and fungicides. R.d 1.66.

Aspect. The longitudinal positions of two celestial objects relative to one another. See conjunction, opposition, quadrature.

Asphalt. Brownish-black substance used in road making, roofing, and waterproofing. A naturally occurring mixture of Hydrocarbons, it is commercially obtained as a residue in the Distillation or refining of Petroleum. Asphalt varies in consistency from a solid to a semisolid, melts when heated, and has great tenacity. Used in paints and varnishes, it imparts an intense black color. Crushed asphalt rock, a natural mixture of asphalt, sand, and limestone, is used as road-building material.

Asphodel. Hardy, stemless herbs (genera *Asphodelus* and *Asphodeline)* of the Lily family, native to India and the Mediterranean. Both have showy flower spikes. Other asphodels include the false asphodel (genus *Tofieldia)* and the mountain asphodel, or turkeybeard *(Xerophyllum asphodeliodes).*

Aspirin. Acetyl derivative of salicylic acid, commonly used to lower fever, relieve headache, muscle, and joint pain; and reduce inflammation, particularly that caused by rheumatic fever and arthritis. Known side effects are nausea, vomiting, diarrhea, and gastrointestinal bleeding. Acetaminophen (Tylenol) is often substituted for aspirin. *See also* Analgesic.

Ass. Hoofed, herbivorous Mammal (genus *Equus),* related to but smaller than, the Horse. Unlike the horse, the ass has a large head, long ears, and small hooves. The two living species are the sandy-coloured wild Asian ass *(E. hemonius),* now endangered, and the larger, gray African ass *(E. hemonius),* now endangered, and the larger, gray African ass (*E. asinus*), from which domestic donkeys are descended.

Associative. Denoting an operation that is not affected by the way in which the terms operated on are grouped. Multiplication and addition of integers are associative since $(n + m) + l = n + (m + l)$, etc.

Associative Law. In mathematics, law holding that for a given operation combining three quantities, two at a time, the initial pairing is arbitrary. Addition and multiplication are associative; thus $(a + b) + c = a + (b + c)$ and $(a \times b) \times c = a \times (b \times c)$ for any three members a, b, and c. Substraction and division, however, are not associative.

Astatine (At). Semimetallic radioactive element, discovered in 1931 by Fred Allison and E.J. Murphy. The heaviest known Halogen, it is believed to be similar to iodine in its chemical properties. Astatine-211 (half-life 7.21 hr) is used as a radioactive tracer because it collects in the thyroid gland.

Astatine. Symbol : At. A radioactive element first prepared by bombarding bismuth with alpha particles. Minute amounts of astatine are also found naturally. The most stable isotope. At, has a half-life of 8.3 hours. Little is known about the chemistry of astatine. It belongs to the halogens and seems to be more metallic than iodine in its behaviour. A.N. 85; m.pt. 302°C; b.pt. 337°C.

Aster. Widely distributed, wild and cultivated perennial flowering plants (genus *Aster)* of the Composite family. Most species have white, pink, blue, or purple flowers that bloom in the fall. The China aster *(Callistephus chinensis)*, the common aster of florists and flower gardens, and the golden asters (genus *Chrysopsis)* are in the same family.

Asteroid, planetoid, or minor planet. Small, usually irregularly shaped body orbiting the sun, most often at least partially between the orbits of Mars and Jupiter. Ceres is the largest asteroid (diameter: 470 mi/750 km) and was the first discovered (1801). Of the more than 2,000 asteroids known, most have been discovered photographically; their paths appear as short lines in a time exposure. Asteroids may be fragments of a planet shattered in the remote past; material that failed to condense into a single planet; or material from the nuclei of old comets. The Trojan asteroids revolve in the same orbit as Jupiter, kept by perturbation effects in two groups 60° ahead of and 60° behind Jupiter. The Apollo asteroids cross the earth's orbit; they may be the cause of the earth's several meteorite craters and are a possible location for future mining.

Asteroid (minor planet, planetoid). Any of a large number of a small celestial bodies orbiting the sun, mainly between Mars and Jupiter. The largest of these rocky objects is *Ceres* with a diameter of 685 kilometres but very few asteroids have diameters greater than 100 km. The orbits of some are very eccentric, such as those of *Adonis* and *Icarus,* which approach the path of Mercury. It is estimated that there are about 100,000 asteroids. They may be a remains of a planet that disintegrated or could not be formed because of Jupiter's massive gravitational influence. Such a planet is allowed for by Bode's law.

Asthma. Chronic respiratory disorder characterized by labored breathing and wheezing resulting from obstructed and constricted air passages. Although asthma usually results from an allergic reaction (see Allergy), specific allergens are not always identifiable. Illness and stress may precipitate an attack. For acute attacks Epinephrine injections and oxygen therapy bring immediate relief. Long-term control includes use of bronchodilators, Steroids, breathing exercises, and, if possible, the identification and avoidance of allergens.

Astigmatism. An aberration of lenses and mirrors, in which the curvature is different in two mutually perpendicular planes, so that a single point off the axis of the system is focused as two different images in the form of short lines. The eye can suffer from astigmatism, a defect remedied by the use of toroidal spectacle lenses.

Astrolabe. An ancient astronomical instrument used for plotting angular distances on the celestial sphere, consisting of a graduated disc with an index arm (the *alidade)* that could be moved and sighted. The angular distance between two objects could be measured by lining up on each in turn and measuring the angle formed at the observer's position.

Astrolabe. Instrument of ancient origin that measured altitudes of celestial bodies and determined their positions and motions. It typically consisted of a wooden or metal disk with the circumference marked off in degrees, suspended from an attached ring. Angular distances were determined by sighting with the alidade—a movable pointer pivoted at the disk's center—and taking readings of its position on the graduated circle. Skilled mariners used astrolabes up to the 18th cent. to determine latitude, longitude, and time of day.

Astrology. Form.of Divination based on the theory that movements of the celestial bodies (stars, planets, sun, and moon) influence human affairs and determine events. The Chaldaeans and Assyrians, believing all events to be predetermined, developed a nondeistic system of divination. The spread of astrological practice was arrested by the rise of Christianity, with its emphasis on divine intervention and free will; but in the Renaissance astrology regained popularity, in part due to rekindled interest in science and Astronomy. Christian theologists warred against astrology, and in 1585 Sixtus V condemned it. At the same time the work of Kepler and others undermined astrology's tenets, although the practice has continued. One's horoscope is a map of the heavens at the time of one's birth, showing the positions of the heavenly bodies in the Zodiac.

Astronaut or **Cosmonaut.** Crew member on a U.S. or Soviet manned spaceflight mission. The early astronauts and cosmonauts were generally trained aircraft test pilots; later astronauts and cosmonauts, however, have included scientists and physicians. As far as is possible, all conditions to be encountered in space, e.g., the physiological disorientation arising from weightlessness (see Space Medicine), are simulated in ground training. Using trainers and mock-ups of actual spacecraft, astronauts and cosmonauts rehearse every maneuver from liftoff to recovery; every conceivable mal-function and difficulty is anticipated and prepared for. Prominent Soviet cosmonauts include Yari Gagarin, Valentina Tereshkova, and Aleksei Leonov. Prominent U.S. astronauts include Alan Shepard, John Glenn, Edward White, Neil Armstrong, Buzz Aldrin, and John Young.

Astronomical Coordinate System. Four basic systems used to indicate the positions of celestial bodies on the celestial sphere. The latter is an imaginary sphere that has the observer at its center and all other celestial bodies imagined as located on its inside surface.

Equatorial. The celestial equator is the projection of the earth's Equator onto the celestial sphere, and the celestial poles are the points where the earth's axis, if extended, intersects the celestial sphere. *Right ascension* (R.A.) is the angle (in hours, minutes, and seconds, with 1 h = 15°) measured eastward from the vernal Equinox to the point where the hour circle (the great circle passing through the celestial poles and the body) intersects the celestial equator. *Declination* is the angle (in degrees, minutes, and seconds, as are all other angles defined hereafter) measured north (+) or sough (–) along the body's hour circle between the celestial equator and the body.

Horizon or Altazimuth. Altitude is the angle measured above (+) or below (–) the observer's celestial horizon (the great circle on the celestial sphere midway between the points—zenith and nadir—directly above and below the observer) to the body along the vertical circle passing through the body and the zenith. *Azimuth* is the angle measured along the celestial horizon from the observer's celestial meridian (the vertical circle passing through the nearest celestial pole and the zenith) to the point where the body's vertical circle intersects the horizon. The earth's rotation constantly changes a body's altitude and azimuth for an observer.

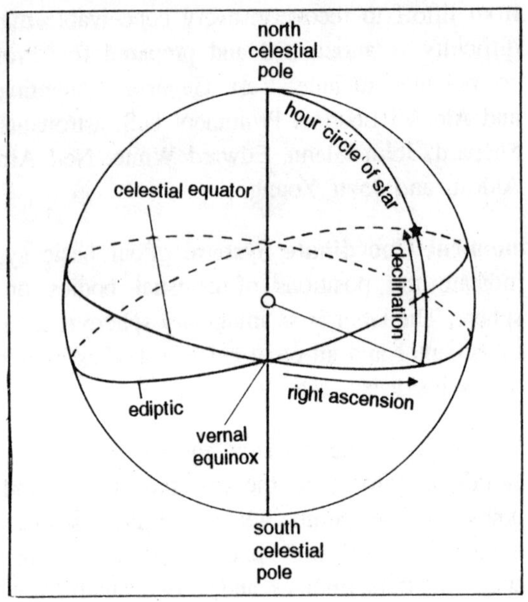

Fig. A-19. Equatorial Astronomical Coordinate System

Ecliptic or Celestial. The ecliptic poles are the two points at which a line perpendicular to the plane of the Ecliptic and passing through the earth's center intersects the celestial sphere. *Celestial latitude* is the angle measured north (+) or south (—) from the ecliptic to the body along the attitude circle through the body and the ecliptic poles. *Celestial longitude* is the angle measured along the ecliptic from the vernal equinox to the latitude circle, in the same sense as R.A.

Galactic. The galactic equator is the intersection of the mean plane of our galaxy with the celestial sphere, and the galactic poles are the two points at which a line perpendicular to the mean galactic plane and passing through the earth's center intersects the celestial sphere. *Galactic latitude* is the angle measured north (+) or south (–) from the galactic equator to the body along the great circle passing through the body and the galactic poles. *Galactic longitude* is the angle measured eastward along the galactic equator from the galactic center.

Astronomical unit. (AU), Mean distance between the earth and the sun. One AU is c. 92,960,000 mi (149,604,970 km).

Astronomy. Branch of science that studies the motions and natures of celestial bodies, such as Planets, Stars, and Galaxies; more generally, the study of matter and energy in the universe at large. Astronomy is perhaps the oldest of the pure sciences. In many primitive civilizations the regularity of celestial motions was recognized, and attempts were made to keep records and predict events. Astronomical observations provided a basis for the Calendar by determining the units of Month and Year. Later, astronomy served in navigation and timekeeping. The earliest astronomers were priests, and no attempt was made to separate astronomy from the pseudoscience of Astrology. Astronomy reached its highest development in the ancient world with the Greeks of the Alexandrian school in the Hellenistic period. Aristarchus of Samos determined the sizes and distances of the Moon and Sun and advocated a heliocentric (sun-centered) cosmology. Eratosthenes made the first accurate measurement of the actual size of the earth. The greatest astronomer of antiquity, Hipparchus, devised a geocentric system of cycles and epicycles (as compounding of circular motions) to account for the movements of the sun and moon. Using such a system, Ptolemy predicted the motions of the planets with considerable accuracy (see Ptolemaic system). One of the landmarks of the scientific revolution of the 16th and 17th cent. was Nicholas Copernicus' revival (1543) of the heliocentric theory (see Copernican System). The next great astronomer, Tycho Brahe, complied (1576-97) the most accurate and complete astronomical observations yet produced. Johannes Kepler's study of Brahe's observations led him to the three laws of planetary motion that bear his name. Galileo Galilei, the first to make astronomical use of the Telescope, provided persuasive evidence (e.g., his discovery of the four largest moons of Jupiter and the phases of Venus) for the Copernican cosmology. Sir Isaac Newton, possibly the greatest scientific genius of all time, succeeded in uniting the sciences of astronomy and Physics. His laws of motion and theory of universal Gravitation, published in 1687, provided a physical, dynamic basis for the merely descriptive laws of Kepler. By the early 19th cent. the science of Celestial Mechanics had reached a highly developed state through the work of Alexis Clairaut, Jean d'Alembert, Leonhard Euler, Joseph Lagrange, Pierre Laplace, and others. In 1838, Friedrich Bessel made the first measurement of the distance to a star (*see* Parallax). Astronomy was revolutionized in the second half of the 19th cent. by techniques based on photography and the Spectroscope. Interest shifted from determining the positions and distances of stars to studying their physical composition (see Stellar Evolution). With the construction of ever more powerful telescopes (*see* Observatory), the boundaries of the known universe constantly increased. Harlow Shapley determined the size and shape of our galaxy, the Milky Way. Edwin Hubble's study of the distant galaxies led him to conclude that the universe is expanding (see Hubble's Law). Various rival theories of the origin and overall structure of the universe, e.g., the big bang and steady state theories, were formulated. Most recently, the frontiers of astronomy have been expanded by Space Exploration and observations in new parts of the spectrum, e.g., gamma-ray astronomy, Radio Astronomy, ultraviolet astronomy, and X-ray astronomy. The new observational techniques have led to the discovery of strange new astronomical objects, e.g., Pulsars, Quasars, and black holes.

Asymmetric carbon atom. *See* optical isomerism.

Asymptote. A line to which a curve approaches as its limit as the value of either the function or its arguments tend to infinity or to some other value for which the function is not defined. More generally, one curve may be an asymptote of another if the difference between the functions tends to zero.

Atactic. *See* polymer.

Athermancy. The property of being opaque to infrared radiation. *Compare* diathermancy.

Atmolysis. The process of separating gases by their different rates of diffusion. If two gases are allowed to diffuse through a porous barrier, the one with lower molecular weight will pass through faster.

Atmosphere. The mixture of gases and other substances surrounding a celestial body with sufficient gravity to maintain it. Although some details about the atmospheres of the other planets and some satellites are known (see articles on individual planets), a complete description is available only for the earth's atmosphere, the study of which is called Meteorology. The gaseous constituents of the earth's atmosphere are not chemically combined, and thus each retains its own physical and chemical properties. Within the first 40 to 50 mi (64 to 80 km) above the earth, the mixture is of uniform composition (except for a high concentration of Ozone at 30 mi/50 km). This whole region contains more than 99% of the total mass of the earth's atmosphere. Based on their relative volumes, the gaseous constituents are nitrogen (78.09%), oxygen (20.95%), argon (0.93%), carbon dioxide (0.03%), and minute traces of neon, helium, methane, krypton, hydrogen, xenon, and ozone. Additional atmospheric constituents include water vapor and particulate matter, such as various forms of dust and industrial pollutants. The earth's atmosphere is separated into certain distinct regions, each having a different temperature range. The troposphere, where air is in constant motion (*see* Wind), extends from the earth's surface to an altitude of 5 mi (8 km) at the poles and 10 mi (16 km) at the equator. Clouds and other Weather phenomena occur here (see also Climate).

All forms of the earth's animal and plant life exist in the troposphere or in the water beneath it. Above the troposphere, the stratosphere extends to about 30 mi (50 km), followed by the mesosphere, up to about 50 mi (80km), the thermosphere, up to about 400 mi (640 km) , and finally the exosphere. The ionosphere is in the range (50 to 400 mi/80 to 640 km) that contains a high concentration of electrically charged particles (ions), which are responsible for reflecting radio signals. Above it, out to about 40,000 mi (64,400 km) in a region called the magnetosphere, electrically charged particles are trapped by the earth's magnetic field (see Aurora, Van Allen Radiation Belts). The atmosphere protects the earth by absorbing and scattering harmful radiation and causing extraterrestrial solid matter to burn from the heat generated by air friction.

Atmosphere. 1. The region of gas surrounding a planet or other heavenly body. The earth's atmosphere is divided into layers depending on temperature gradient or electrical properties. The composition of dry air, by volume, is given in the table.

Atmosphere

Composition of air by volume

nitrogen	78.08%
oxygen	20.94%
argon	0.933%
carbon dioxide	0.03%
neon	0.0018%
helium	0.0005%
krypton	0.0001%
xenon	0.000009%

In addition, air contains a variable amount of water vapour, dust particles, and usually small quantities of hydrocarbons, sulphur compounds, and other pollutants.

2. Symbol: atm. A unit of pressure equal to 101 325 pascals. Before 1954 the atmosphere was taken to be 760 millimetres of mercury.

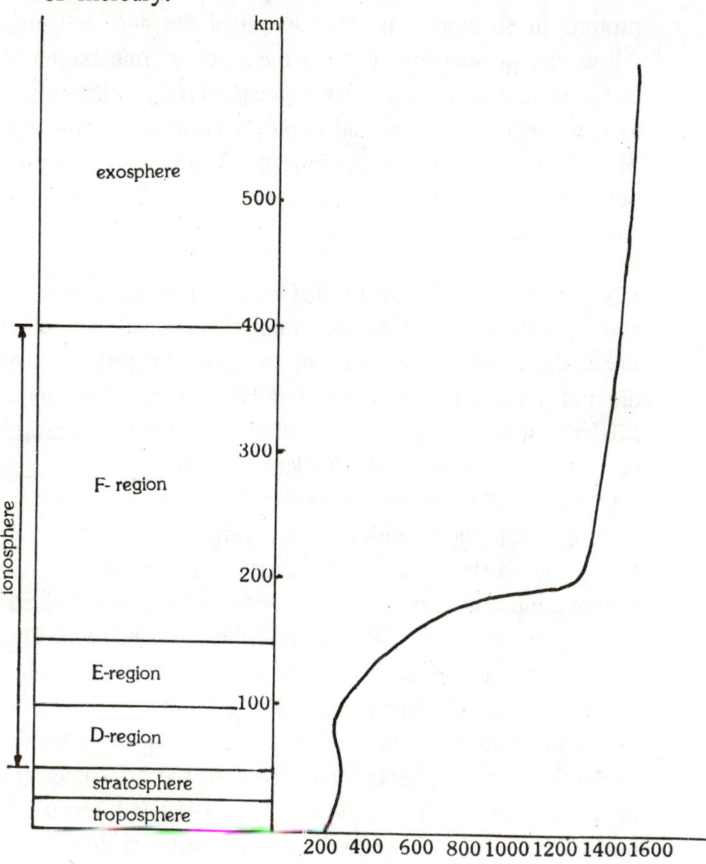

Fig. A-20. Atmospheric layers.

Atom. The smallest unit of a chemical Element having the properties of that element. An atom contains several kinds of particles. Its central core, the nucleus, consists of positively charged particles, called Protons, and uncharged particles, called Neutrons. Surrounding the nucleus and orbiting it are negatively charged particles, called Electrons. Each atom has an equal number of protons and electrons. The nucleus occupies only a tiny fraction of an atom's volume but contains almost all of its mass. Electrons in the outermost orbits determine the atom's chemical and electrical properties. The number of protons in an atom's nucleus is called the Atomic Number. All atoms of an element have the same atomic number and differ in atomic number from atoms of other elements. The total number of protons and neutrons combined is the atom's Mass Number. Atoms containing the same number of protons but different numbers of neutrons are different forms, or Isotopes, of the same element.

History. In the 5th cent. B.C. the Greek philosophers Democritus and Leucippus proposed that matter was made up of tiny, indivisible particles in constant motion. Aristotle, however, did not accept the theory, and it was ignored for centuries. Modern atomic theory began with John Dalton who proposed (1808) that all atoms of an element have exactly the same size and weight, and that atoms of elements unite chemically in simple numerical ratios to form compounds. In 1911 Ernest Rutherford explained an atom's structure in terms of a positively charged nucleus surrounded by negatively charged electrons orbiting around it. In 1913 Niels Bohr used Quantum Theory to explain why electrons could remain in certain allowed orbits without radiating energy. The development of quantum mechanics during the 1920s resulted in a satisfactory explanation of all phenomena related to the role of electrons in atoms and of all aspects of their associated spectra. The quantum theory has shown that all particles have certain wave properties. As a result, electrons in an atom cannot be pictured as localized in space but rather should be viewed as a cloud of charge spread out over the entire orbit. The electron clouds around the nucleus represent regions in which the electrons are most likely to be found. Physicists are currently studying the behaviour of large groups of atoms, and the nature of and relations among the hundreds of Elementary Particles that have been discovered in addition to the proton, neutron, and electron.

Atom. One of the large number of particles that together make up a given element. An atom is the smallest amount of a chemical element that can take part in a chemical reaction. All atoms of the same element are identical. Atoms were originally thought of as small rigid indivisible particles (see Dalton's atomic theory). They are now known to be made up of simpler particles: protons, neutrons, and electrons. The atomic radius is of the order of 10^{-10} metre but almost all its mass is concentrated in a small central region — the nucleus — with a radius of 10^{-15} metre. The nucleus has a positive charge and is composed of protons and, usually, neutrons. Negatively charged electrons move around this central region, the number of electrons being equal to the number of protons so that the atom is electrically neutral. An atom of a given element has a characteristic number of protons (or electrons) — this is the *atomic number*. Thus, hydrogen atoms have one proton, helium atoms have two lithium three, etc. The element may also have different isotopes, containing different numbers of neutrons in their nuclei.

In early models of the atom the electrons were thought of as moving in fixed circular or elliptical paths around the nucleus (*see* Bohr atom). The modern idea, resulting from wave mechanics, is that no precise path can be given. Instead, our knowledge of the location of electrons is based on probability and each electron is assigned to a region in space (*see* atomic orbital).

Whichever model is used, the orbital electrons in atoms occupy certain energy levels characterized by quantum numbers. First, the electrons are grouped into shells denoted by letters K, L, M, N, etc. These are characterized by the *principal quantum number (n)*, which may have values 1, 2, 3, 4, etc. Each shell can contain a maximum of $2n^2$ electrons: thus the K shell is full when it contains 2 electrons, the L shell has 8, the M shell 18 etc. The shells have sub-shells characterized by an *azimuthal quantum number (l)*. This gives the orbital angular momentum of the electron and has values 0, 1, 2, ... (n - 1). The first four values are denoted by letters *s, p, d,* and *f* respectively. The K shell has only one sub-shell (i.e. the sub-shell is the same as the shell). This is an *s* sub-shell. The L shell has two sub-shells; an *s* sub-shell and *p* sub-shell. Within any given type of sub-shell, there may be more than one energy level (or atomic orbital). These are characterized by a *magnetic quantum number (m)* with values - l, (-l - 1), ... 0 ...(l - 1), If 1 - 0 (an *s*-orbital), m = 0. If l - 1 (a *p*-orbital), m = -1, 0, and l; i.e. there are three *p*-orbitals within this sub-shell. Usually these are of identical energy but their energies differ in the presence of an external magnetic field. Similarly there are five *d*- orbitals and seven *f*-orbitals. Finally, an electron in an atom is characterized by a fourth quantum number describing its spin; this can have values of either + 1/2 or - 1/2. Each level (or orbital) can contain a maximum of two electrons with opposing spins.

The pattern of energy levels given by the four quantum numbers determines the configuration of the atom of a particular element and explains the groupings of elements in the periodic table.

Atomic bomb. Weapon deriving its great explosive force from the sudden release of Nuclear Energy through the fission, or splitting, of heavy atomic nuclei. The first atomic bomb was successfully tested by the U.S. near Alamogordo, N. Mex., on July 16, 1945 (*see* Manhattan Project). In the final stages of World War II the U.S. dropped atomic bombs on Hiroshima on Aug. 6, 1945, and on Nagasaki three days later to force Japan to surrender. Atomic bombs were subsequently developed by the USSR (1949), Great Britain (1952), France (1960), China (1964), and India (1974). Practical fissionable nuclei for atomic bombs are the isotopes Uranium-235 and Plutonium-239, which are capable of undergoing chain reaction. If the mass of the fissionable material exceeds the critical mass, the chain reaction multiplies rapidly into an uncontrollable release of energy. An atomic bomb is detonated by bringing together very rapidly (e.g., by means of a chemical explosion) two subcritical masses of fissionable material. The ensuring explosion produces great amounts of heat, a shock wave, and intense neutron and gamma radiation. The region of the explosion becomes radioactively contaminated, and wind-borne radioactive products may be deposited elsewhere as fallout.

Atomic energy. Energy obtained by nuclear fission or fusion.

Atomic heat. The heat capacity of one mole of a substance.

Atomicity. The number of atoms in a molecule of a given element. Carbon dioxide, for example, has an atomicity of three.

Atomic mass unit (dalton). Symbol: amu. A unit of mass equal to one twelfth of the mass of a neutral carbon 12 atom. It is equivalent to 1.660 33 X 10^{-27} kilogram. This is the unit on the international scale. Formerly, atomic mass unit were defined in terms of the mass of the oxygen atom. The physical atomic mass unit was one sixteenth of the mass of the neutral oxygen-16 atom. The chemical atomic mass unit was one sixteenth of the weighed average of the masses of the three naturally occurring isotopes of oxygen. These units were slightly different: 1 amu (international) is equal to 1,000 31 amu (physical) or to 1,000 04 amu (chemical). The atomic mass unit is not to be confused with the atomic unit of mass (see Hartree units).

Atomic number (proton number). Symbol. Z. The number of protons in the nucleus of a given atom. The atomic number is also the number of electrons in the atom, thus determining the chemical properties of the element.

Atomic Orbital. A region around the nucleus of an atom in which an electron moves. According to wave mechanics, the electron's location is not described by a fixed orbit but by a probability distribution in space, given by the wave function in Fact, there is a finite probability of finding the electron at any point: the orbitals are regions within which the electron may be found with a high degree of probability.

Each orbital has a fixed energy and contains a maximum of two electrons. It is characterized by three quantum numbers *n, l* and *m* (*see* atom). The shape depends on the value of *l*.

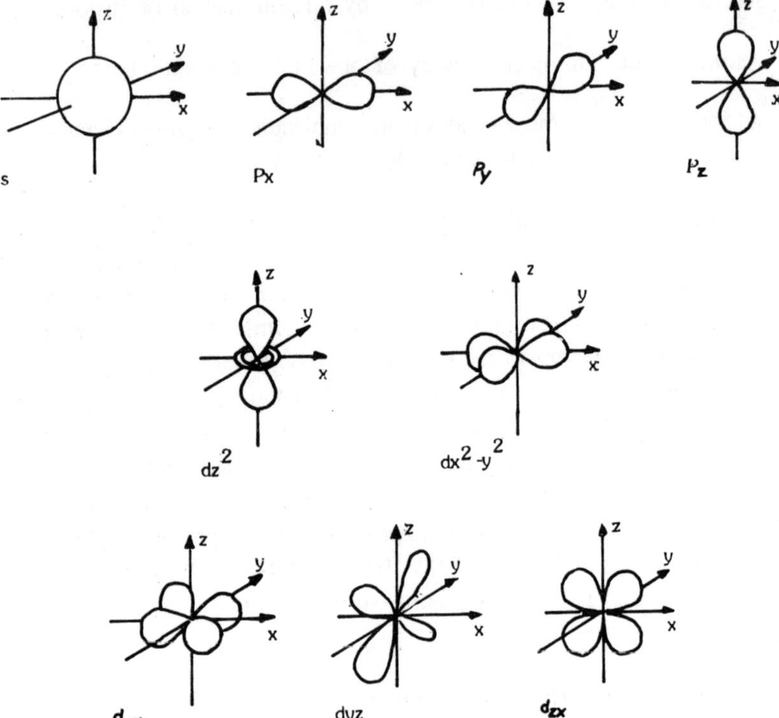

Fig. A-21. Shapes of atomic orbitals.

Atomic Weight. Mean (weighted average) of the masses of all the naturally occurring Isotopes of a chemical Element; the atomic mass is the mass of any individual isotope. Atomic weight is usually expressed in atomic mass units (amu); the atomic mass unit is defined as exactly 1/12 the mass of a carbon-12 atom. Each proton or neutron weighs about 1 amu, and thus the atomic mass is always very close to the mass number (total number of protons and neutrons in the nucleus). Because most naturally occurring elements have one principal isotope and only insignificant amounts of other isotopes, most atomic weights are also very nearly whole numbers. For the atomic weight of individual elements.

Atropine. A poisonous white crystalline alkaloid, $C_{17}H_{23}NO_3$, obtained from deadly nightshade and similar plants. It is used in medicine. M.pt. 114-6°C.

Attenuation. Decrease in intensity of a signal, electromagnetic wave, sound wave, beam of particles, etc.

Atto-. Symbol: a. Prefix indicating 10^{-18}; 1 attometre (am) = 10^{-18} metre.

Audio frequency. A frequency in the range 20 to 20,000 hertz, this being the range in which the human ear can detect sound.

Audiovisual instruction. The use of nonverbal as well as verbal materials, particularly pictures and sounds, to promote learning. Audiovisual devices, formerly limited to static materials such as maps, graphs, and textbook illustrations, were used successfully as instructional tools by the U.S. armed forces during World War II. As technology developed, audiovisual instruction began to include films, photographs, sound and video recordings, and television, as well as Programmed Instruction provided through computers and other types of teaching machines. Instructional television is widely used in schools, industry, and the military; with the use of cable television and electronic audio and video equipment, it can be available in the home as well.

Auger effect. The phenomenon in which an excited ion decays to the ground state and ejects an electron, producing a doubly charged ion. An excited ion has a vacancy in one of its inner electron shells and this can be filled by transition of an electron from an outer shell. The resulting release of energy can be taken up by an emitted photon, as in X-ray fluorescence. The Auger effect is an alternative process in which the energy is used to ionize the singly charged ion. [After Pierre Auger (b. 1899), French physicist].

Auk. Swimming and diving Bird of the family Alcidae, which includes the Puffin and guillemot. Clumsy on land, auks seldom leave the water except to nest; they return to the same nesting site every year. The largest species, the flightless great auk *(Pinguinus impennis)*, was hunted for its flesh, feathers, and oil; it became extinct c. 1844.

Aurora. A display of luminous arcs, curtains, and streamers of light, ever-changing in shape and intensity and varying in colour from whitish-green to deep red, seen in the sky in the regions of the poles. Apparently linked to solar flares, the aurorae are caused by charged particles reaching the earth where they are trapped by the earth's magnetic field. They spiral towards the poles where they caused ionization of molecules in the atmosphere, thus producing light. In the north the effect is called the *aurora Borealis,* while in the south it is referred to as the *aurora Australis*.

Austenite. A solid solution of carbon or iron carbide in gamma iron. It is produced in some steels by quenching from a high temperature.

Autocatalysis. A type of catalysis in which the catalyst is a product of the reaction being catalysed. The reaction starts slowly, then speeds up more and more as the amount of catalyst builds up.

Autoclave. A thick-walled vessel, usually of steel, designed for carrying our chemical reactions at temperatures and pressures.

Autoimmune disease. General term for several disorders in which the body produces antibodies (see Immunity) against its own substances, resulting in tissue injury. For example, in systemic lupus erythematosus, individuals develop antibodies to their own nucleic acids and cell structures, causing dysfunction of many organs, including the heart, kidneys, and joints. Autoimmune diseases are treated by a variety of non-specific Immunosuppressive Drugs and Steroids.

Automation. Automatic operation and control of machinery or processes by devices that make and execute decisions without human intervention. Such devices use self-correcting control systems that employ feedback; i.e., they use part of their output to control their input. Because of their ability to store, select, record, and present data, computers are widely used to direct automated systems.

Automobile. Self-propelled vehicle used for travel on land. The fundamental structure of the automobile consists of seven basic systems. The engine usually mounted in front and driving either the two front or the two back wheels; the fuel system, using a Carburetor to produce the optimal combustible mixture of fuel and air, the electrical system, including a battery that provides a power source to the ignition; the cooling, steering and suspension, and brake systems; and the Transmission, which transmits power from the engine crankshaft to the wheels by means of a series of gears. Evolving from earlier experiments with steam-powered vehicles, models using the gasoline-fueled internal-Combustion Engine were first developed by the German engineers Karl Benz (1885) and Gottlieb Daimler (1886) U.S. leadership in automobile production began with Henry Ford's founding (1903) of the Ford Motor Co., its production (1908) of the inexpensive Model T, and its development of assembly-line techniques. General Motors, Ford's principal competitor, became the world's largest automobile manufacturer in the 1920s, and U.S. dominance of the field continued until the 1970s, when it was challenged by growing sales of Japanese and German cars.

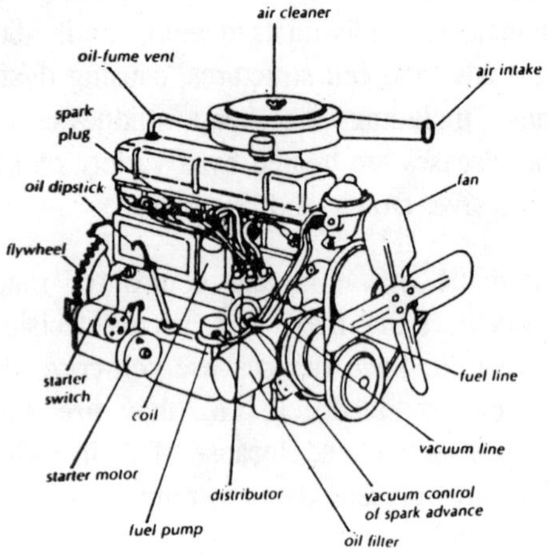

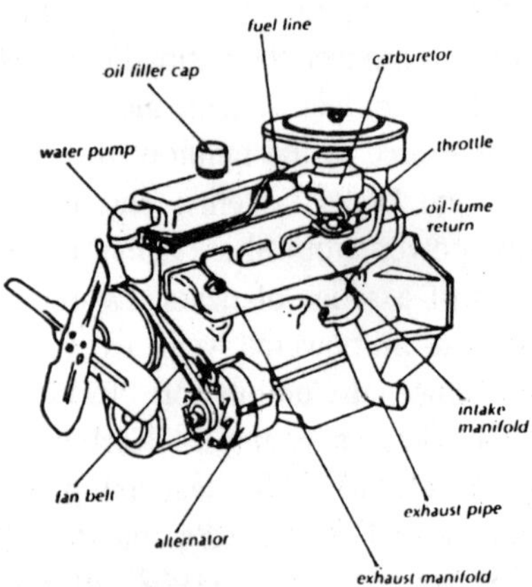

Fig. A-22. Automobile: Two views of a six-cylinder, gasoline-fueled engine.

Avalanche. Rapidly descending mass of snow and ice loosened from a mountain slope. Sudden and often destructive, avalanches result from the addition of a heavy snowfall to an insecure mass of ice and snow, from the melting or erosion of part of the base of the mass, or from sudden, shocks such as those caused by explosions or earth tremors.

Aviation. Operation of heavier-than-air aircraft and related activities. Aviation can be divided into military aviation air transport (commercial airline operations), and general aviation (agricultural, business charter, instructional, and pleasure flying). The first successful flights of a motor-powered airplane carrying a human were made by Orville and Wilbur Wright near Kitty Hawk, N.C. on Dec. 17, 1903. The first successful Seaplane was constructed in 1911-12. During the early 1900s, aviators demonstrated the feasibility of air travel to various parts of the world. World War I provided additional motivation for aviation research and development. The availability of cheap, surplus aircraft in the U.S. after the war encouraged barnstorming and stunt-flying; the result was a more airplane-conscious public. Private companies in America contracted the carrying of airmail after 1925. Technological improvements in Wind Tunnel testing, engine and airframe design, and maintenance equipment combined in the 1930s to provide faster, larger, and more durable airplanes. The transportation of passengers became profitable, and routes were extended to include several foreign countries. Transpacific airmail service began in 1934, and was soon followed by a similar service for passengers. In 1939 the first transatlantic service for mail and passengers was inaugurated. The application of Jet Propulsion to commercial air transportation began in 1952. The first supersonic transports (SST) for passenger service were put into service during the mid-1970s.

Avocado. Tropical American broad-leaved evergreen tree (genus *Persea)* of the Laurel family, and its pear-shaped fruit. The fruit has a tough, inedible, usually dark green skin and an oily flesh surrounding a large, hard seed. The flesh is eaten fresh, chiefly in salads.

Avogadro, Amadeo, conte di Quaregna. 1776-1856, Italian physicist. In 1811 he advanced the hypothesis (since known as Avogadro's law) that equal volumes of gases under identical conditions of pressure and temperature contain the same number of molecules. This hypothesis led to the determination by other physicists of the value of Avogadro's number, i.e., the number of molecules in one mole, or gram-molecular weight, of any gas.

Avogadro constant. Symbol : L or N_A. The number of molecules in one mole of any substance. It has the value 6.022 52 x 10^{23}. [After Amedeo Avogadro (1776-1856), Italian physcist].

Avogadro's Hypothesis. The principle that equal volumes of all gases at the same temperature and pressure contain the same number of molecules.

Avoirdupois weight. The system of weights based on the pound.

Axiom. In Mathematics and Logic, general statement accepted without proof as the basis for logically deducing other statements (Theorems). Examples of exioms used widely a mathematics are those related to equality (e.g., "If equals are added to equals, the sums are equal") and those related to operations (e.g., the Associative law). A postulate, like an axiom, is a statement that is accepted without proof; it deals, however, with specific subject matter (e.g., properties of geometrical figures), not general statements.

Azalea. Shrubs (genus *Rhododendron)* of the Heath family, distinguished by typically deciduous leaves and large clusters of pink, red, orange, yellow, purple, or white flowers. Most grow in damp acid soils of hills and mountains, and are native to North America and Asia. Native American azaleas include the flame azalea *(R. calendulacea)* and the fragrant white azalea *(R. viscosa),* also called swamp honey-suckle. Most of the brilliantly flowered garden varieties are from China and Japan.

Azeotropic mixture (azeotrope). A mixture of two liquids in such proportions that it boils without change in composition; i.e., the composition of the vapour is the same as that of the liquid. In general, when a mixture is distilled the more volatile constituent vaporizes first. When the composition reaches that of the azeotropic (or *constant-boiling)* mixture, the constituents distil together. Ethanol and water, for instance, have an azeotropic mixture containing 4.4% of water: further separation cannot be effected by simple distillation.

Azide. 1. A Salt of hydrazoic acid, containing the ion N_3. The azides of heavy metals are often explosive and are used as detonators.

2. An organic compound with the formula RN_3 where R is a hydrocarbon group.

Azimuth. The angular distance of a vertical circle passing through a celestial object and the observer's zenith from the south point of the observer's horizon. The azimuth is measured westward from the south point, which is taken as 0°.

B

B. Chemical symbol of the element Boron.

Ba. Chemical symbol of the element Barium.

Baade, Walter. German-American astronomer. He presented evidence for the existence of two different Stellar Populations of older and new stars. Baade knew that, at the then-accepted distance of the Andromeda galaxy, cluster-type variable stars should have appeared on photographs that he took with the 200-in. (5.08-m) telescope at Palomar Observatory. Because they did not, he correctly reasoned (1952) that the distances to this galaxy and other extragalactic systems must be doubled.

Babbage, Charles, 1792-1871. English mathematician and inventor, famous for his attempts to develop a mechanical computational aid he called the "analytical engine." Although it was never constructed and was decimal rather than binary in conception, it clearly anticipated the modern digital Computer. A scientist with extremely broad interests, Babbage probed the roles of learned societies and government in advancing science and wrote on mass production and on what is now called operational research.

Babbitt metal. Any of various alloys of tin (80-90%), antimony (7-11%), and copper (4-8%), used in bearings.

Baboon. Large, powerful ground-living Monkey (genus *Papio*), also called dog-faced monkey, related to the Mandrill. Found in the open country of Africa and Asia, baboons have close-set eyes under heavy brow ridges, long, heavy muzzles, powerful jaws, cheek pouches for storing food, and sharp, tusklike upper canine teeth. Baboons have a highly developed social structure.

Babo's law. The principle that the vapour pressure of a liquid is lowered when a solid is dissolved in it, by an amount proportional to the concentration of solid dissolved.

Baby's breath. Name for a plant of the Pink family, for the white bedstraw of the Madder family, and for grape hyacinth of the Lily family.

Back e.m.f. An electromotive force opposing the flow of a current in a circuit. In cells a back e.m.f., reducing the cell's natural e.m.f., can be produced by polarization. Back e.m.f.s can also arise by induction. In a circuit in which the current is changing, the back e.m.f. is caused by self-induction. Similarly, in a conductor moving in a magnetic field, as in an electric motor, there is a back e.m.f. induced by the changing flux. Induced e.m.f.s are produced according to Lenz's law.

Backgammon. Game of chance and skill played by two persons on a specially marked board divided by a space (bar) into two "table," each of which has 12 alternately colored points (elongated triangular spaces) on which each player places 15 pieces (disks) in a prescribed formation. Two dice are thrown to determine moves. The object is to be the first to move one's pieces around and off the board. The game has very ancient roots in the Middle East.

Background. A residual signal in a counter or other detector of radiation observed in the absence of the radiation that is measured. The background signal is caused by cosmic rays, small amounts of radioactive material in the surroundings, etc.

Bacteria. Microscopic, unicellular organisms having three typical forms: rod-shaped (bacillus), round (coccus), and spiral (spirillum). The cytoplasm of most bacteria is surrounded by a cell wall, the nucleus contains DNA but lacks the nuclear membrane found in higher plants and animals. Many forms are motile, propelled by movements of a filamentlike appendage (flagellum). Reproduction is chiefly by transverse fission (Mitosis), but conjugation (transfer of nucleic acid between two cells) and other forms of genetic recombination also occur. Some bacteria (aerobes) can grow only in the presence of free of atmospheric oxygen; others (anaerobes) cannot grow in its presence; and a third group (faculative anaerobes) can grow with or without it. In unfavorable conditions, many species from resistant spores. Different types of bacteria are capable of innumerable chemical metabolic transformations, e.g., Photosynthesis and the conversion of free nitrogen and sulfur into Amino Acids. Bacteria are both useful and harmful to humans. Some are used for soil enrichment with leguminous plants, in pickling, and alcohol and cheese fermentation, to decompose organic wastes (in septic tanks and the soil), and in Genetic Engineering. Others, called pathogenes, cause a number of plant and animal diseases, including Cholera, Syphilis, Typhoid Fever, and Tetanus.

Bacteriophage or **Phage.** Virus that infects Bacteria, sometimes destroying them. A phage has a head composed of Protein and an inner core of nucleic Acid. It infects a host by attaching itself by its tail to the bacterial cell wall and injecting nucleic acid (DNA) into its host, in which it produces new phage particles. Eventually the bacterial cell is destroyed by lysis, or dissolution, releasing the phage particles to infect other cells. Phages are highly specific, with a particular phage infecting only certain species of bacteria; they are important tools in studies of bacterial genetics and cellular mechanisms.

Baddedeyite. A black, brown or yellowish naturally occuring oxide of zirconium, ZrO_2.

Badger. Any of several related members of the Weasel family. Most are large, nocturnal burrowers with broad, heavy bodies, long snouts, sharp claws, and long, grizzled fur. The Old World badger (*Meles meles*), found in Europe and N. Asia, weighs about 30 lb (13.6 kg) and feeds on rodents, insects, and plants. The smaller American badger (*Taxidea taxus*) has short legs and a white stripe over the forehead and around each eye; a swift burrower, it will pursue prey into their holes and may construct its own living quarters 30 ft (9.1 m) below ground level.

Bagpipe. Musical instrument, most widely used in Ireland and Scotland, consisting of an inflated bag, usually leather; one or two chanters (or chaunters) melody pipes with fingers holes; and one or more drones; which produce one tone each. The bagpipe is an ancient instrument, probably carried E and W from Mesopotamia by Celtic migrations.

Bailey's beads. The phenomenon characterized as a necklace effect surrounding the dark body of the moon just before and after totality during a total solar eclipse. It is caused by sunlight catching the irregular lunar surface features and shining into valleys on the moon's limb.

Bakelite. O Any of various synthetic phenon-formaldehyde resins.

Bakelite. [For its inventor, Leo Baekeland], a synthetic thermosetting phenon-formaldehyde Resin with an unusually wide variety of industrial applications ranging from billiard balls to electrical insulation.

Baking Powder. Any mixture used in baking as a substitute for yeast in generating carbon dioxide and thus causing the dough to rise. The commonest type is a mixture of sodium bicarbonate and either tartaric acid or cream of tartar.

Baking Soda. *See* sodium bicarbonate.

Balance. An instrument for measuring weight, consisting of a rigid beam balanced on a fulcrum at its midpoint with a pan suspended from each end. The object to be weighed is placed in one pan and standard weights added to the other until the beam is horizontal. The accuracy of sensitive balances is increased by use of a *rider*—a small weight moved along the top of the balance beam along a calibrated scale. Laboratory balances have agate knife edges to reduce friction and are enclosed in a case to reduce errors from air currents and temperature changes. The beam has a vertical pointer attached to its mid-point to enable the balance position to be detected. Such instruments are capable of measuring to 0.001 g. Other instruments for weighing use elastic deformation, as in the spring balance.

Bald Cypress. Common name for the Taxodiacea, a family of deciduous or evergreen Conifers with needlelike or scale-like leaves and woody cones. Most species are native to the Far East; some e.g., the big trees and redwood and bald cypresses, are native to North America. Almost all are cultivated for ornament. The common bald cypress (*Taxodium distichum*), valued for its softwood, forms dense forest in the SE U.S. and is common in the Everglades. Bald cypresses are called "bald" because of their deciduous character, unusual in conifers. True Cypresses belong to a separate family.

Ballistic galvanometer. A galvanometer with a relatively long period of oscillation, for use in measuring an electric charge. The charge is allowed to flow quickly through the galvanometer. If the natural period of oscillation is long compared with the time for which the current flows, the current passes before the coil or magnet begins to move. In the case of a moving-coil instrument the initial deflection (or "throw") is proportional to the charge.

Ballistic pendulum. A heavy pendulum used to measure the velocity of a fast projectile such as a bullet. It consists of a heavy freely suspended mass which the projectile hits—the momentum, and thus the velocity, being obtained from the subsequent deflection of the pendulum.

Ballistics. Science of projectiles, such as bullets, bombs, rockets, and missiles. Interior ballistics deals with the propulsion and motion of a projectile while in flight, and includes the study of the trajectory, or curved flight path, of the projectile. Terminal ballistics is concerned with the phenomena occuring at the termination of the projectile's flight; such termination may result from impact on a solid target or explosion of the projectile. In criminology the term *ballistics* is applied to the identification of the weapon from which a bullet was fired. Microscopic imperfections in a gun barrel make characteristic scratches and grooves on bullets fired though it.

Balloon. Lighter-than-air craft without a propulsion system, lifted by inflation of one or more containers with a gas lighter than air or with heated air. During flight, altitude is gained by discarding ballast, e.g. bags of sand, and lost by releasing some of the lifting gas from its container. In some late designs using air heated by a gas-fired burner, the altitude is controlled by varying the temperature of the heated air. The balloon was invented by the French brothers Joseph and Jacques Etienne Montgolfier, who in 1783 caused a linen bag about 100 ft (30 m) in diameter to rise in the air. Using a Montgolfier balloon, Pilatre de Rozier and the marquis d'Arlanders made the first manned balloon flight on Nov. 21, 1783. The Americans Ben Abruzzo, Maxie Anderson, and Larry Newman made (1978) the first successful transatlantic balloon crossing. Today Weather Balloons equipped with radio transmitters and other instruments transmit meteorological readings to ground stations at regular intervals. High-altitude balloons are used in astronomy, especially in the study of cosmic rays and the photography of other planets.

Balsam (oleoresin). Any of various substances consisting of a mixture of a natural resin and an essential oil, exuded from certain trees and shrubs. Balsams are semi-solids or viscous liquids, usually with pleasant aromatic odours. They are used in medicine, perfumery, varnishes, and lacquers.

Bamboo. Plant (genus *Bambusa*) of the Grass family, chiefly of warm or tropical regions. The genus contains the largest grasses, sometimes reaching 100 ft (30 m). Bamboo stalks are hollow, usually round, and jointed, with deciduous leaves. Bamboo is used as wood, and for construction work, furniture, utensils, fiber, paper, fuel, and innumerable other articles. Bamboo sprouts and the grains of some species are eaten. Native American bamboo is a Cane

Banana. Name for a family of tropical herbs (the Musacae), for a genus (*Musa*) of herbaceous plants and for the fruits they produce. Bananas are probably native to tropical Asia, but are widely cultivated. They are related to the economically valuable Manila Hemp and to the Bird-of-Paradise Flower. Banana plants have palmlike aspects and large leaves, the overlapping bases of which form the so-called false trunk. Only female flowers develop into the banana fruit (botanically, a berry), each plant bearing fruit only once. The seeds are sterile; propagation is through shoots from the rhizomes. Bananas are an important food staple in the tropics.

Band Spectrum. An emission or absorption spectrum consisting of a series of bands, each of which is formed of a number of closely spaced lines. Band spectra are produced by molecules. Each band corresponds to a transition between electronic energy levels and the lines within bands result from different vibrational energy levels of the molecule.

Band Theory (of solids). The application of quantum mechanics to the energies of electrons in crystalline solids. In isolated atoms the orbiting electrons can have only certain fixed energies: i.e. they occupy discrete energy levels. In solids, these levels become bands of allowed energy made up of closely-spaced energy levels, each capable of containing a pair of electrons. The allowed energy bands are separated by bands of forbidden energy.

The band theory of solids explains differences in the electrical conductivity of different materials. When a solid conducts electricity, electrons move through the lattice under the influence of an applied electric field. According to quantum mechanics, they can only gain energy from the field in difference between two energy levels. Thus in accelerating they make transitions from one energy level to a higher level. This can only occur if the energy band is not completely fitted by electrons: i.e. if there are higher vacant energy levels available, the situation occuring in metals and other good conductors. In electrical insulators the bands are filled and there are no vacant allowed energy levels for the electrons to occupy.

Semiconductors have filled energy bands, in which all the possible energy levels are occupied, but the forbidden band is narrow. In intrinsic semiconductors it is possible for electrons to gain enough thermal energy to move into a higher vacant band. The electrons excited to this band can then conduct electricity and further conduction occurs as a result of holes left in the lower energy band. The conductivity increases with temperature. Extrinsic semiconductors owe their properties to the presence of energy bands due to impurities, lying in the forbidden gap between the filled band and the higher empty band. Impurities that accept electrons from the filled band create positive holes and produce p-type semiconductivity. Electron donors supply electrons to the empty band and cause n-type semiconductivity.

Bandwidth. 1. The range of frequencies over which an amplifier or other electronic device operates. The band-width is taken to be the range over which the device performs at some specified fraction of its maximum performance.

2. The frequency range over which most of the power is transmitted in a radio transmission. The bandwidth is defined with reference to a specified fraction of the total power, usually 99%.

Baobab, huge tree (*Adansonia digitata*) of the Bombax family, native to India and Africa. The trunk diameter of this relatively short tree is exceeded only by that of the Sequoia, and the trunks of living trees are hollowed out for dwellings. The bark is used for rope and cloth, the leaves yield condiments and medicines, and the gordlike fruit (monkeybread) is eaten.

Bar. A unit of pressure equal to 10^5 pascals. The bar was formerly defined as the pressure resulting from a force of one dyne acting on an area of 1 square centimetre.

Barbiturate. Any depresant drug derived from barbituric acid. In low doses, barbiturates have a tranquilizing effect. Increased doses are hypnotic or sleep-including, and still larger doses act as anticonvulsants and anesthetics. Barbiturates are commonly taken as sleeping pills; such use may lead to psychological dependency, physiological tolerance, and even death by overdose. Barbiturates do not relieve pain.

Bardeen, John, 1908-, American physicist; b. Madison, Wis. He is known for his studies of semiconductivity and other aspects of Solid-state Physics. The first to win a Nobel Prize twice in the same field, Bardeen shared the 1956 physics prize with Walter Brattain and William Shockley, for work in developing the Transistor, and the 1972 physics prize with Leon Cooper and John Schreiffer, for their theory of Super-conductivity.

Barff process. A process for protecting iron against corrosion by heating the hot metal in steam, thus forming a protective layer of black iron oxide: $2Fe + 3H_2O = Fe_2O_3 + 3H_2$ [After F. S. Barff, 19th-century U.S. engineer.]

Barite (barytes, heavy spar). A white or colourless mineral consisting of barium sulphate, $BaSO_4$.

Barium. Symbol: Ba. A silvery-white fairly malleable metallic element occurring principally in barite ($BaSO_4$) and witherite ($BaCo_3$). Produced by electrolysis of the fused chloride, it is used in some alloys and as a getter. Barium is one of the alkaline-earth metals. A.N. 56; A.W. 137.34; m.pt. 850°C; b.pt. 1140°C; r.d. 3.78; valency 2.

Barium carbonate. A poisonous heavy white powder, $BaCO_3$, used in rat poison, optical glass ceramics, paints, and enamels. Barium carbonate occurs in nature as the mineral witherite. Decomposes at 1300°; r.d.4.27.

Barium chloride. A colourless poisonous crystalline salt, $BaCl_2.2H_2O$, used as an additive for lubricating oils. M.pt. 960°C (anhydrous); r.d. 3.10.

Barium hydroxide. A white alkaline hydroxide, $Ba(OH)_2$, obtained as a crystalline solid or powder by adding water to barium oxide. The monohydrate, pentahydrate, and octahydrate can also be made. Barium hydroxide is used in sugar refining and in the manufacture of glass. M.pt. 78°C; r.d.2.13.

Barium oxide. A white or yellowish deliquescent basic oxide, **BaO,** obtained as a powder by fusion of barium sulphate with carbon in an electric furnace. It is used as a dehydrating agent. M.pt. 1920°C; r.d.5.72.

Barium peroxide. A grey-white powder, BaO_2, made by heating barium oxide in a stream of oxygen. The compound is used in the manufacture of hydrogen peroxide. M.pt. 450°; r.d.4.96.

Barium sulphate. A white heavy insoluble salt, $BaSO_4$, occurring naturally as barite. It is used in paint pigments and in X-ray photography. M.pt. 1580°C; r.d.4.25-4.5.

Bark, outer covering of the Stem of woody plants, composed of waterproof Cork cells (the outer bark) protecting a layer of food-conducting tissue (the inner bark or phloem). As the stem grows in size (see Cambium), the outer bark gives way by splitting, shredding, or peeling in patterns typical of the species. Various barks are sources of textile fibres (e.g., Hemp, Flax, and Jute), tannin, cork, dyes, flavorings (e.g., Cinnamon), and drugs (e.g., Quinine).

Barkhausen effect. The effect observed when a ferromagnetic material is magnetized, in which the magnetization does not increase continuously but in a series of small steps. These are interpreted as the alignment of magnetic domains. The Barkhausen effect can be demonstrated by winding two coils on an iron core. As the current in one is slowly increased, fluctuations, detectable by an earphone or oscilloscope, occur in the other. [After Heinich Barkhausen (1881-1956), German physicist.]

Barley. Annual cereal plant (*Hordeum vulgare*) of the Grass family, widely cultivated except in hot and humid climates Barley, was known to the ancients was the chief bread grains in Europe as late as the 16th cent. Today, each of the many varieties grown has a special purpose, e.g., for stock feed for malting, as a minor source of flour, and in soups.

Barn. Symbol: b. A unit of area equal to 10^{-28} sq. metre. It is used for expressing the cross section of atomic nuclei in scattering experiments.

Barnacle, sedentary marine animal (subclass Cirripedia), a Crustacean. Barnacles permanently attach themselves to a substrate by means of an adhesive cement. They secrete a calcareous shell around themselves and form conspicuous encrusting colonies on rocks, pilings, boats, and some marine animals (e.g., whales, turtles). The attached end of the animal is the head; jointed legs (cirri) sweep food particles through the shell opening to the mouth. Some barnacles lack shells and are Parasites of other invertebrates.

Barograph. A barometer that produces a continuous record of pressure with time: usually an aneroid barometer with the pointer carrying a pen that traces a curve on a moving chart.

Barometer, 1. Instrument for measuring atmospheric pressure. The mercurial barometer consists of a mercury-filled glass tube that is sealed at one end and inverted in a cup of mercury. Pressure on the surface of the mercury in the cup supports the mercury in the tube, which varies in height depending on variations in atmospheric pressure. At 32°F (0°C), standard sea-level pressure (1 standard atmosphere) is 14.7 lb/in.^{2} (1,030 g/cm^2), which is equivalent to a column of mercury 29.92 in. (76 cm) in height. The aneroid barometer contains sealed, partially evacuated metallic box. As the air pressure on it varies, one of its surfaces expands or contracts; this motion is transmitted by a train of levers to a pointer, which shows the pressure on a graduated scale. In Weather forecasting, a rising barometer usually indicates fair weather; a rapidly falling barometer, stormy weather.

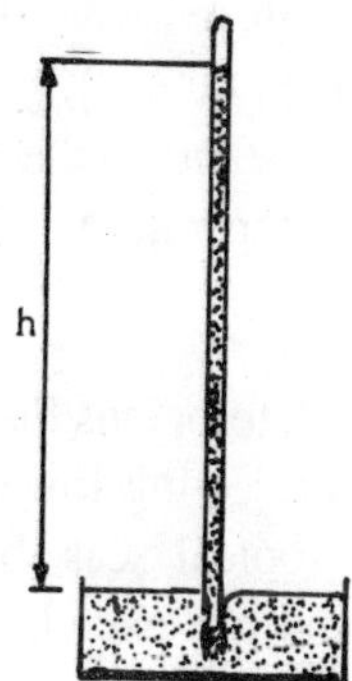

Fig. B-1. Mercury barometer.

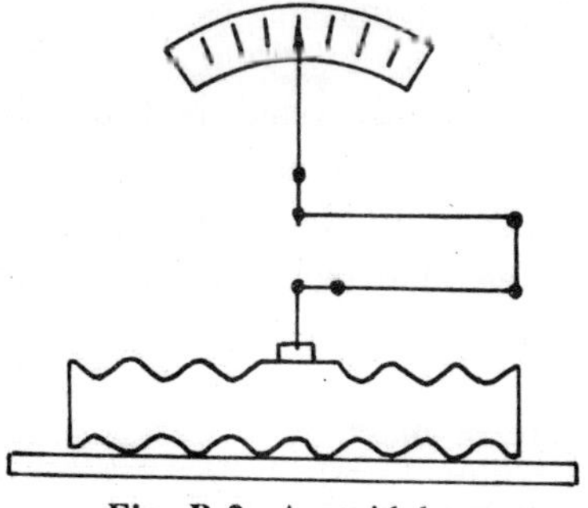

Fig. B-2. Aneroid barometer.

2. An instrument for measuring atmospheric pressure. The

Mercury barometer consists of a long glass tube sealed at one end. It is filled with mercury and inverted in an open reservoir of mercury. The atmospheric pressure is the vertical height of mercury that the atmosphere will support (normally 760 mm).

The *aneroid barometer* contains a sealed metal box from which all the air has been removed. It has a corrugated top of very thin metal attached to a pointer by a train of levers. Changes in atmospheric pressure cause movements in the corrugated top, thus causing the pointer to move.

Barometric formula. The formula $p = p_o \exp(-mgx/kT)$, relating the pressure p to the height x above the earth's surface. p_o is the pressure at the surface ($x = 0$), g the acceleration of free fall, k the Boltzmann constant, m the mass per molecule, and T the thermodynamic temperature, which is assumed to be constant.

Barracuda. Slender, elongated, ferocious Fish (family Sphyraenidae) with a long snout and projecting lower jaw edged with large, sharp teeth. Found in tropical seas, barracudas will strike at anything that gleams and are excellent game fish. The great barracuda (*Sphyraena barracuda*), up to 10 ft (3 m) in length, is dangerous to swimmers.

Barrel distortion. *See* distortion.

Barrow. In archaeology, burial mound, built usually of earth and stone or timber, perhaps to simulate cave burial. In W. Europe long barrows date from the New Stone Age and round barrow from the Bronze Age. More recent Asian round barrow (Stupas) usually house Buddhist relics.

Barycentre. *See* centre of mass.

Baryon. Any of a class of elementary particles that have half-integral spin and take part in strong interactions. They generally have a mass larger than that of the mesons. Baryons are fermions: they include the proton, the neutron, and the hyperons. It is possible to assign a quantum number (the *baryon number, B*) to each particle, defined to be 1 for the baryons and -1 for their antiparticles. Particles that are not baryons have a baryon number 0. In any reaction the total baryon number is unchanged. *Compare* lepton.

Baryta. *See* barium oxide.

Barytes. *See* barite.

Basalt. Fine-grained igneous Rock of volcanic origin, with a high percentage of iron and magnesium, in a range of dark colours. Its texture varies depending on conditions of cooling. Most of the world's great Lava flows (e.g. the Columbia Plateau in the NW U.S.) are basalt. It underlies the sediment cover in the world's oceans and is believed to underlie the Continents as well. Many of the lunar rocks obtained by the Apollo astronauts are basalt.

Base. 1. A substance that will react with an acid to form a salt and water. Bases turn litmus blue and produce hydroxide ions if dissolved in water. They are usually oxides or hydroxides of metals or solutions of ammonia or other amines. *Strong bases* are substances that are completely dissociated into ions in solution. Sodium hydroxide is an example : it consists of a mixture of Na^+ and OH^- ions. *Weak* bases are partially dissociated. Ammonia, for example, reacts with water to give ammonium ions and hydroxide ions: $NH_3 + H_2O = NH_4^+ + OH$. Soluble bases are alkalis. The ideas of a base in chemistry has been extended to include Lewis bases.

2. The region of a transistor between the emitter and the collector.

3. The horizontal straight line at the bottom of a geometric figure, or the bottom face of a geometric solid.

4. The number of units grouped together in a positional number system and represented by 1 in the next position. Thus decimal numbers have a base of ten: binary numbers have a base of two.

Base metal. A common metal of relatively low value, such as iron or lead, as distinguished from noble metals such as gold or other rare or precious metals.

Basic. Denoting a compound that is a base or a solution that contains an excess of hydroxide ions.

Basic dye. Any of a class of dyes that consist of salts containing large coloured positive ions. Basic dyes can be used on acidic acrylic fibres and as direct dyes for wool and silk. They can also be used on cotton with a tannic acid mordant. Triphenylmethane derivatives are examples of basic dyes.

Basic salt. *See* salt.

Basic slag. A by-product in the manufacture of steel, consisting of a mixture of calcium phosphate, calcium silicate, lime, and ferric oxide. It is used as fertilizer for its high phosphorus content.

Basil. Tender herb or small shrub (genus *Ocimum*) of the Mint family, cultivated for the aromatic leaves. Common, or sweet, basil (*O. basilicum*) is used for seasoning. Holy basil and bush basil are related plants.

Basketry. Art of weaving or coiling flexible materials to form utensils for the preparation, storage, or serving of food; boats; huts; traps; apparel; and other objects. Twigs, roots hide, bamboo, cane raffia, grass, or straw are used. In antiquity, Egyptians (4000-5000) B.C.) used baskets for storing grain, and North American Indians of the Southwest (c. 1500 B.C.) covered baskets with clay and baked them to create fireproof cooking vessels, anticipating pottery. In some parts of the world pottery preceded basketry.

Bass. Any of various Fishes of the families Serranidae (sea basses) and Centrachidae (black basses and sunfishes). Sea basses, a large, diverse family of fishes with oblong, rather compressed bodies, inhabit warm and temperate seas worldwide and are highly valued as food and game fish. The largest sea basses are Groupers. Sunfished, spiny-finned, freshwater fishes with flattened bodies, are found in North America. Black basses, averaging 2 to 3 (.9 to 1.4 kg), are the most valuable American freshwater game fishes.

Bat. The only Mammal (order Chiroptera) capable of true flight. Numbering between 1,000 and 2,000 species, bats range in size from less than 1 in. (2.5 cm) to 15 in. (45 cm), with a wingspan of from less than 2 in. (5 cm) to 5 ft (150 cm). The body is furry and mouselike, with the forelimbs and extensions of the skin of the back and belly modified to form wings. Bats are most abundant in the tropics, and temperate species often hibernate or migrate to warmer areas in the winter. Most species frequent crevices, caves, or buildings, and are active at night or twilight; they roost during the day, often in large number and usually hanging by their feet. Most bats see well but depend on echolocation to navigate in the dark. Bats are fruit-eaters (fruit, nectar, pollen) or insect-eaters (fruit, insects, small animals, and fish); one species, the South American vampire bat, feeds exclusively on the blood of living animals, chiefly mammals.

Batholith. Enormous mass of igneous Rock, granitic in composition, with steep walls and without visible floors. Batholiths commonly extend over thousands of square miles. Their method of formation is a controversial subject; most appear to have moved up through the earth's crust in a molten state, shattering and incorporating the overlying country rock.

Battery. An electric cell or a number of cells connected together. The car battery is a lead-plate accumulator. The common dry batteries used in torches, radios, etc., are Leclanche cells.

Batik. Method of decorating fabric used for centuries in indonesia. With melted wax, a design is applied to the cloth (cotton or, sometimes, silk), which is then dipped in cool vegetable dye. Areas covered by wax do not receive the dye and display a light pattern on the coloured ground. The process may be repeated several times. When the design is complete, the wax is removed in hot water. A crackling effect occur if dye has seeped into cracks of hardened wax. The same or similar patterns have been used for c.1,000 years. Batik was brought to Europe by Dutch traders, was adopted in the 19th cent. by Western craftsmen, and is still widely used.

Bauxite. A naturally occurring hydrated aluminium oxide, $Al_2O_3nH_2O$, containing ferric oxide and silica. It is the most important ore of aluminium.

Bazooka. Military weapon consisting of a portable, light-weight tube that serves as a rocket launcher, usually operated by two men. Developed by the U.S. as an Infantry weapon for use against Tanks, pillboxes, and bunkers, it was widely employed during World War II and the Korean War but was later superseded by more powerful, accurate weapons, notably recoilless weapons and antitank missiles.

Beadle, George Wells, 1903-. American geneticist; b. Wahoo, Nebr. For their work on the bread mold *Neurospora crassa,* which showed that genes control cellular production of enzymes and thus the basic chemistry of cells, Beadle and Edward Tatum shared the 1958 Noble Prize in physiology or medicine with Joshua Lederberg.

Beagle. Small, compact Hound, shoulder height, 10-15 in. (25.4-39.1 cm); weight, 20-40 lb (9.1-18.1 kg). Its short, close-lying, harsh coat is usually black, tan, and white. Developed in England to hunt hares, it was introduced into the U.S. in the 1870s.

Bean. Name for seeds of trees and shrubs of the Pulse family, and for various plants of the family having edible seeds or seed pods. True beans are in the genus *Phaseolus*. The cowpea, Carob, and CHick-Pea and sometimes considered beans. Cultivated worldwide, beans are an important food staple. They are high in protein and often used as a meat substitute.

Bear. Large Mammal of the family Ursidae, found almost exclusively in the Northern Hemisphere. Bears have large heads, bulky bodies, short, powerful, clawed limbs, and coarse, thick fur; almost all are omnivorous. In cold climates, bears do not hibernate but sleep most of the winter; their metabolism remains normal, and they may wake and emerge during warm spells. North American brown bears include the Kodiak bear and the girzzly. The Kodiak, the largest bear, sometimes stands 9 ft (2.7 m) high and weighs over 1,600 lb (730 kg). The grizzly, characterized by silver-tripped (grizzled) fur, is more highly carnivorous than most bears and preys on large mammals such as deer. The most widespread North American bear is the black bear, smaller than the brown bears and ranging in colour from light brown to black. The large, white polar bear is an arctic species. A solitary, fearless hunter, it is a powerful swimmer and feeds mostly on marine animals.

Bearing. Machine part used to reduce Friction between moving surfaces and to support moving loads. There are two principal types. A *plain,* or *journal, bearing* is a cylinder that supports a rotating shaft such as a motor shaft; its inner lining, the bushing, is usually made of a metal softer than that of the shaft, so that any slight misalignment of the shaft can be adjusted by an equivalent wearing of the bushing. An *antifriction bearing* is a cylinder containing a movable inner ring of small steel balls (the ball bearings used primarily in light machiner) or larger cylindrical rollers. The rotating machine part fits into a center of the ring, which takes up the motion of the rotating part, distributing and reducing friction through the movement of its bearings.

Beats. Regular fluctuations in the intensity of sound heard when two tones of nearly equal frequency are sounded together. The beats result from interference between the tones and have a frequency, the *beat frequency*, given by the difference in the frequencies of the two tones.

Beaver. Large aquatic Rodent *(Castor fiber)*. Once widespread in the Northern Hemisphere, beavers are from 3 to 4 ft (91 to 120 cm) long, including the distinctive broad, flattened tail; they usually weigh about 60lb (27kg). Known for their engineering feats, beavers create ponds by building dams of sticks, logs, and mud; they build habitations, or lodges, in the same way.

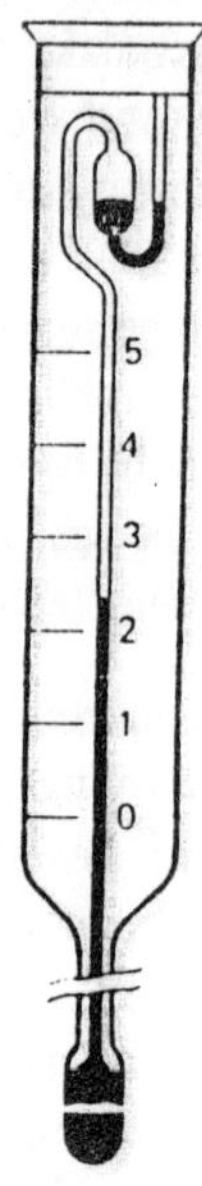

Fig. B-3. Beckmann thermometer.

Beckmann thermometer. A mercury thermometer designed for measuring small changes of temperature. The capillary tube has a scale of about 5°C and leads into a reservoir at the top of the thermometer (see illustration). The instrument can be used near any temperature in the range 0-100°C by setting the amount of mercury in the bottom bulb using the reservoir. [After Ernst Beckmann (1853-1923), German physical chemist.]

Bedbug. Small, flatbodied, blood-sucking Bug of the family Cimicidae. Distributed worldwide, bedbugs are parasites of warm-blooded animals. They are reddish-brown and about 1/4 in. (6 mm) long.

Bee. Flying insect of the superfamily Apoidae, having enlarged hind feet for pollen gathering and a dense coat of feathery hairs on the head and thorax. Bees feed on Pollen and nectar; the latter is converted to Honey in the digestive tract. Most have stings connected to a poison gland. Bees may be social, solitary, or parasitic in the nests of other bees. Social bees include bumblebees, stingless bees, and honeybees. A typical colony of social bees has an egg-laying queen, sexually undeveloped females (workers), and fertile males (drones). Workers gather nectar, make and store honey, and protect the hive. They care for the queen and larvae and perform complex patterned dances to communicate the location of pollen sources to one another. After being fertilized by a drone, the queen spends her life (usually several years) laying eggs. Honeybees are raised commercially for honey and for the Wax they produce for their nests (combs). Bees are invaluable for cross-Pollination.

Beech. Common name for the Fagaceae, a family of trees and shrubs mainly of temperate and subtropical regions in the Northern Hemisphere. The principal genera, Chestnut, chinquapin, beech, and Oak, are dominant forest trees valued for their hardwood timber. Some species are also grown for their fruits and as ornamentals. Th beeches have smooth,. silvery gray bark and pale green leaves. The American beech *(F. grandifolia)* grows over much of the NE U.S. and Canada. The European beech *(F. sylvatica)* is an important forest tree, prized for its wood and nut oil. Several of its varieties, e.g., purple and copper beeches, are cultivated in America as ornamental trees.

Beer. One of the oldest known alcoholic beverages. At first brewed chiefly in the household and monastery, it has been a commercial product since late medieval times and is made today in most industrialized countries. Color, flavor, and alcoholic content (usually 3%-6%) may vary, but the process is brewing is similar: A mash of malt (usually barley), cereal adjunct (e.g., rice and corn), and water is heated and agitated. The liquid is boiled with hops and cooled. Yeast is then added, and fermentation occurs. In England ale is any light-colored beer; in the U.S. it is a pale, strongly hopped beverage. Porter is a strong, dark ale brewed with roasted malt; stout is darker, stronger, and maltier than porter. Bock beer is dark, heavy, and usually drunk in the spring. Light, or low-calorie, beer is lower in alcohol content.

Beet. Biennial or annual root vegetable (*Beta vulgaris*) of the goosefoot family, cultivated since pre-Christian times. Numerous varieties exist, e.g., red, or garden, beet; sugar beet; and Swiss chard. Both the roots and foliage of the red beet are edible, as is the foliage of the Swiss chard. The widely cultivated sugar beet, containing up to 20% sucrose, provides about one third of the world's sugar supply.

Beetle. Mainly terrestrial Insect (order Coleoptera). Beetles have chewing mouthparts; well-developed antennae; and a pair of hard, opaque, waterproof wings (elytra) which cover and protect the flight wings and the body. Some species are brilliantly colored and patterned, but most are dull. They range in size from less than 1/32 in. (1 mm) to more than 6 in. (15 cm) in length. Beetles are generally plant eaters, but some are parasitic.

Behaviorism. School of psychology seeking to explain behavior entirely in terms of observable responses to environmental stimuli. Influenced by the conditioned-reflex experiments of Pavlov, behaviorism was introduced in 1913 by J.B. Watson who, denying both the value of introspection and the concept of consciousness, emphasized laboratory techniques. B.F.

Skinner, the major modern proponent, has concerned himself exclusively with the relationship of observable responses to stimuli and rewards.

Behaviour Modification. Treatment of human behavioral disorders through the reinforcement of acceptable behavior and suppression (by non-reinforcement) of undesirable behavior. Practiced by means of various techniques of reward and/or punishment (e.g., aversion therapy, desensitization), it has been used to treat such disorders as Drug Addiction, Alcoholism, and Phobias. A form of Psychotherapy developed from the work of B.F. Skinner and other, behaviour modification is employed in private and institutional therapy in both group and individual settings. The method has been criticized by some for treating symptoms rather than causes.

Bel. *See* Decibel.

Bell. In music, a Percussion Instrument consisting of a hollow metal vessel set into vibration by a blow from a clapper within or a hammer without. Apparently originating in Asia, bells have been used in connection with all major religions except Islam. A portable set of bells tuned to the intervals of the major scale is called a chime. A carillon is a larger stationary set of up to 70 bells with chromatic intervals played from a keyboard. It developed in the Low Countries and reached a peak there in the 17th cent. while in England the art of Change Ringing evolved. Carillon playing declined at the end of the 18th cent., but was revived in the 20th cent. following the rediscovery of older tuning secrets and the development of improvements in construction.

Belladonna or **Deadly Nightshade.** Poisonous perennial plant (*Atropa belladona*) of the Nightshade family. Native to Europe and now wild in the U.S., the plant has reddish, bell-shaped flowers and shining black berries. Extracts of the leaves and roots dilate the pupils of the eye and were once so used by women—hence the name *belladonna,* meáning "beautiful lady" in Italian. Belladonna has also been employed as a poison, a sedative, and, in medieval times, a Hallucinogenic Drug. The Alkaloid drug atropine, an extract of belladonna, is frequently used to relax muscles and suppress glandular and mucous secretions.

Bell Metal. Any alloy of copper (about (80%) and tin, sometimes with small amounts of zinc and lead.

Benedict Solution. A blue aqueous solution of sodium carbonate, copper sulphate, and sodium citrate. It is used as a test for reducing agents, such as glucose, in the presence of which it forms a red-to-yellow precipitate. [After Stanley Benedict (1884-1936), U.S. Biochemist.]

Benzal Chloride. *See* Benzylidene Chloride.

Benzaldehyde. A yellowish fragrant volatile oily aldehyde, C_6H_5CHO, occurring naturally in the kernels of bitter almonds. It is made by the oxidation of toluene and used in the manufacture of dyes, perfumes, and flavours.

Benzanilide. A white or reddish crystalline powder, $C_6H_5CONHC_6H_5$, manufactured from benzoic anhydride and aniline. Benzanilide is used in the synthesis of drugs, dyes, and perfumes.

Benzene. A clear colourless highly flammable liquid, C_6H_6, with a characteristic odour. The simplest aromatic hydrocarbon, it is obtained by catalytic reforming in the refining of petroleum and from coal tar. Products manufactured using benzene include styrene, detergents, nylon, and insecticides.

Benzene Problem. *See* Aromaticity.

Benzenesulphonic Acid. A colourless crystalline sulphonic acid, $C_6H_5SO_2OH$, made by sulphonation of benzene and used mainly in organic synthesis.

Benzidine. A white, grey, or reddish crystalline powder,

$NH_2(C_6H_4)_2NH_2$, produced by the reduction of nitrobenzene with zinc dust and used in the manufacture of dyes and as an analytical reagent.

Benzil. A yellow crystalline solid, $C_6H_5COCOC_6H_5$, prepared from benzoin by oxidation with nitric acid and used in organic synthesis.

Benzoate. Any salt or ester of benzoic acid.

Benzoic Acid. A white crystalline carboxylic acid, C_6H_5COOH, occurring naturally in many plants and in benzoin gum. It is manufactured by the catalytic decarboxylation of phthalic anhydride and is used in curing tobacco and preserving food.

Benzoin. A white or yellowish crystalline optically active ketone, $C_6H_5CH\text{-}(OH)(COC_6H_5)$ made by the condensation of benzaldehyde in potassium cyanide solution. It is used in organic synthesis.

Benzoquinone. *See* quinone.

Benzoyl Chloride. A colourless pungent liquid acid chloride, C_6H_5COCl, used as a benzoylating agent.

Benzoyl group. The monovalent group, C_6H_5CO-.

Benzoyl Peroxide. A white crystalline explosive solid, $(C_6H_5CO)_2O_2$, made by reaching benzoyl chloride with sodium peroxide and used chiefly as a bleaching agent and a catalyst for free-radical reactions.

Benzyl Alcohol. A colourless liquid, $C_6H_5CH_2OH$, used as a solvent and, in the form of its esters, in perfumery.

Benzyl Group. The monovalent group $C_6H_5CH_2$–.

Benzylidine Chloride (Benzal Chloride). A colourless liquid, $C_6H_5CHCl_2$, used in making benzaldehyde.

Benzidine Group. The divalent group $C_6H_5CH=$, derived from toluene.

Fig. B-4. Benzyne.

Benzyne. A short-lived intermediate, C_6H_4, present in certain reactions.

Bergius Process. A process for manufacturing oil from coal by mixing the coal with heavy oil to form a paste that is mixed with a catalyst and heated under hydrogen at high pressure (about 250 atmospheres). A mixture of liquid hydrocarbons is produced. [After Friedrich Bergius (1884-1949), German chemist].

Beriberi. Disease caused by a deficiency of thiamine (vitamin B_1), resulting in neurological and gastrointestinal disturbances. See table under vitamin.

Berkelium. Symbol: Bk. A transuranic actinide element made, in trace amounts, by neutron bombardment of curium. The most stable isotope, ^{249}Bk, has a half-life of about 300 days. A.N. 97; valency 3 or 4.

Bermuda Triangle. Area in the Atlantic Ocean off Florida where a number of ships and aircraft have vanished. Also known as the Devil's Triangle, it is bounded at its points by Melbourne, Fla.; Bermuda; and Puerto Rico. Storms are common in the region, and investigations to date have not produced scientific evidence of any unusual phenomena involved in the disappearances.

Bernoulli. Swiss family distinguished in scientific and mathematical history. *Jakob, Jacques,* or *James Bernoulli,* 1654-1705, was one of the chief developers of both the ordinary calculus and the calculus of variations. His *Ars conjectandi* (1713) was an important treatise on the theory of probability. His brother, *Johann, Jean,* or *John Bernoulli,* 1667-1748, was famous for his work on integral and exponential calculus; he was also a founder of the calculus of variations and contributed to the study of geodesics, complex numbers, and trigonometery. His son, *Daniel Bernoulli,* 1700-1782, has often been called the first mathematical physicist. His greatest work was his *Hydrodynamica* (1738), which included the principle now known as Bernoulli's principle and anticipated the law of conservation of energy and the kinetic-molecular theory of gases developed a century later. He also made important contributions to probability theory, astronomy, and the theory of differential equations. Other members of the family were noted in the fields of mathematics, physics, astronomy, and geography.

Berthollide Compound. A chemical compound whose atoms are not in simple ratio to each other. Such compounds are said to be nonstoichiometric. An example is titanium oxide, $TiO_{1.8}$, which is deficient in oxygen. [After Claude Louis Berthollet (1748-1822), French chemist.]

Berthelot, Pierre Eugene Marcelin (bertelo). 1827-1907, French chemist. Professor at the Ecole Superieure de Pharmacie and later at the College de France, he became a member of the French Academy in 1900. A founder of modern organic chemistry, he was the first to synthesize organic compounds (e.g., methanol, ethanol, benzene, and acetylene), thereby dispelling the old theory of a vital force inherent in organic compounds. He also worked in thermo-chemistry and in explosives.

Berthollet, Claude Louis, Comte (bertola), 1748-1822, French chemist. Noted for his ideas on chemical affinity and his discovery of the reversibility of reactions, he supported Antoine Lavoisier's theory of combustion and collaborated with him in reforming chemical nomenclature. He analyzed ammonia and prussic acid and discovered the bleaching properties of chlorine.

Beryl. Extremely hard beryllium and aluminum silicate mineral ($Be_3Al_2Si_6O_{18}$), occurring in crystals that may be of enormous size and are usually white, yellow, blue, green, or colorless. It is commonly used as a Gem, the most valued variety being the greenish Emerald; the blue to bluish-green variety is Aquamarine. Beryl is the principal raw material for the element Beryllium and its compounds.

Beryllium (Be). Metallic element, first isolated in 1828 independently by Friedrich Wohler and Antoine Bussy. The silver-gray, Alkaline-Earth Metal is light, strong, high-melting, and resistant to corrosion. It is used as a window material for X-ray tubes and as a shield and a moderator in nuclear reactors.

Berzelius, Jons Jakob, Baron. 1779-1848, Swedish chemist. He developed the modern system of symbols and formulas in chemistry, made a remarkably accurate table of atomic weights, analyzed many chemical compounds, and discovered the elements Selenium, Thorium, and Cerium. He coined the words *isomerism, allotropy,* and *protein.*

Bessel, Friedrich Wilhelm. 1784-1846. German astronomer and mathematician. His discovery of the parallax of the fixed star 61 Cygni, announced in 1838, was the first fully authenticated measurement of a Star's distance from the earth. By 1833 Bessel had increased to 50,000 the number of stars whose positions and proper motions were accurately determined. He established a class of mathematical functions, named for him, as a result of his work on planetary perturbation.

Bessemer Process. Industrial process for the manufacture of Steel from molten Pig Iron. The process is carried out in a large steel container, called the Bessemer converter. iron is introduced through an opening in the narrow upper end. When air is forced upward through perforations in the bottom, impurities such as manganese, silicon, and carbon unite with oxygen to form oxides. The carbon monoxide burns off, and the other impurities form slag. The steel is then poured from the upper opening into molds, and the slag is left behind. Bessemer steel is used to make machinery, tools, wire, and nails, and is the essential modern structural steel used in steel-framework buildings.

Bessemer Process. A process for making steel by blowing air through molten pig iron to oxidize the carbon. The molten iron is contained in a large cylindrical steel vessel (a *Bessemer converter*) and air is introduced through holes in the bottom. Silicon and manganese impurities are also oxidized and removed as slag. The converter is lined with a basic refractory material to remove phosphorus. In some processes a mixture of air and steam is used to prevent absorption of nitrogen by the steel. When the carbon has been removed, the correct quantity of carbon is added to give the type of steel required. [After Sir Henry Bessemer (1813-98), English inventor.]

Beta blocker or **beta adrenergic Blocking Agent.** Drug that reduces the symptoms connected with Hypertension, cardiac arrythmias, Migraine headaches, and other disorders related to the central Nervous System. Within the central nervous system, beta receptors are located mainly in the heart, lungs, kidneys, and blood vessels. Beta blockers compete with Epinephrine for these receptor sites and interfere with the action of epinephrine, lowering blood pressure and heart rate, stopping arrythmias, and preventing migraine headaches. Propranolol is a commonly used beta blocker.

Beta Particle. A particle emitted in one of the three forms of natural Radioactivity.

Beta Decay. A type of radioactive decay in which an unstable nucleus ejects either an electron and an antineutrino or a positron and a neutrino. The product has the same mass number but its atomic number has increased by 1. Tritium, for example, with one proton and two neutrons emits electrons. The process is decay of one neutron to give a proton, an electron, and an antineutrino. The product is helium-3, containing two neutrons and one proton. Nitrogen-13 decays by emission of positrons to give carbon-13. See also radis activity.

Beta Iron. Iron in the temperature range within which it changes from a ferromagnetic material to a paramagnetic material. See also alpha iron.

Beta Particle. An electron, especially one ejected by radioactive decay. Streams of beta particles are called *beta rays.*

Betatron. A type of particle accelerator in which electrons are accelerated by a changing magnetic field. The electrons have a fixed orbit in a torus-shaped evacuated vacuum chamber held between the pole-pieces of a ring-shaped magnet. The magnetic field is pulsed, producing a varying magnetic flux linkage with the orbit and inducing a tangential electric field. The betatron thus works by magnetic induction: an accelerating potential is produced in the same way as the voltage that is induced in a coil of wire placed in a varying magnetic field.

The particle's momentum is proportional to the size of the magnetic field and this can be controlled to increase the electron energies and maintain a stable orbit. High energies (up to 300 MeV) are only possible with light particles, i.e. with electrons. The efficiency is limited by practical constraints on the size of the magnet.

Betel. Masticatory made from seeds of the betel palm (*Areca catechu*). Slices of the seeds (also called betel nuts), together with other aromatic flavorings and lime paste, are smeared onto a betel pepper (*Piper betle*) leaf, which is then rolled up and chewed. Betel contains a narcotic stimulant and has been chewed in S Asia since ancient times.

Bi-. 1. Prefix indicating two. In the names of chemical compounds *di-* is usually used. Thus, carbon bisulphide, CS_2, is more commonly called carbon disulphide.

2. Prefix indicating the presence of hydrogen in an acid salt. The recommended method of naming such compounds is to use *hydrogen* in the name. For example, sodium bisulphate, $NaHSO_4$, is more correctly called sodium hydrogen sulphate.

Biaxial Crystal. A crystal with two optic axes. See double refraction.

Bicarbonate. Any acid salt of carbonic acid. They are more correctly called *hydrogen carbonates*.

Bicycle. Light, two-wheeled vehicle driven by pedals. A model using pedals, cranks, drive rods, and handlebars was introduced c.1839 in Scotland; it was followed by the development of the hollow steel frame, ball bearings, metal wheel spokes, and rubber-rimmed wheels. The front wheel, directly powered by the pedals, was at one time much larger than the rear wheel. The first bicycle with a sprocket-chain drive powering an equal-size rear wheel was made in England in 1885; the pneumatic tire was invented in Scotland in 1888. Later improvements included handlebarmounted brake cables and gearshift systems to facilitate changes in speed. The bicycle's popularity brought about major improvements in roads in 19th-cent. Europe and America, and the vehicle continues to be far more widely used than the automobile in many parts of the world today.

Biennial. Plant requiring two growing seasons to complete its life cycle, as distinguished from an Annual or Perennial. In the first year the plant produces leaves and a fleshy root; in the second year it produces flowers and seeds, and then dies. Some biennials will bloom in the same growing season if sown early.

Big-bang Theory (superdense theory). The theory that the universe began at some definite point in time by a violent explosion of a superdense collection of high-energy matter-a "primordial fire-ball". According to the theory, this matter has continued to fly apart, slowing down as it moves outwards and eventually forming galaxies and stars. Thus the observed expansion of the universe is a result of the "big bang" and the age of the universe can be calculated from red shifts to be 5-50 X 10^9 years.

Evidence for the big-bang model comes from observations of red shifts at far points in the universe. Since the radiation takes many light-years to reach the earth from distant space, such observations give an idea of conditions existing nearer the time of the universe's origin and they show that the universe used to be more dense and was expanding faster. Background radiation has also been discovered that is thought to be a remnant of the big-bang. The main rival to the big-bang theory, the steady-state theory, has less experimental support. The velocity at which the universe is expanding is decreasing. In some modifications of the theory it is predicted that the universe will eventually stop expanding and begin to contract under the influence of gravity until once again it is compressed into a superdense state. This can then expand again—the whole cycle repeating itself indefinitely. Such a model is known as the *oscillating universe.*

Bile. Bitter, alkaline fluid of a yellow, brown, or green color that aids in the digestion of fats (See Digestive System). Composed of water, bile salts, bilirubin (a pigment), cholesterol, and lecithin, bile is secreted by the Liver and stored in the Gall Bladder. It is emptied into the upper intestine to break down fats and enable them (and fat-soluble Vitamins) to be absorbed through the intestinal wall. Bile is also a route of excretion for cholesterol, heme, and many drugs. Jaundice may result if the flow of bile is impeded.

Billion. One million million; 10^{12}. In the U.S. a billion is one thousand million, 10^9.

Bimetallic Strip. A strip consisting of two pieces of different metals welded together so that as the temperature rises the strip bends because of unequal amounts of expansion. Bimetallic strips are used in thermostats and safety devices to break an electric circuit at a preset temperature.

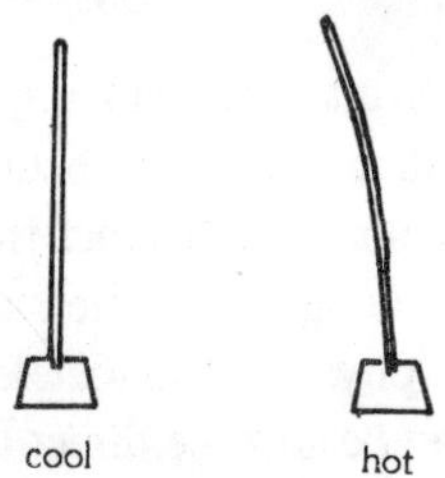

Fig. B-5. Bimetallic strip.

Bimetallism. In economic history, a monetary system in which two commodities, usually gold and silver, were used as a standard and coined at a fixed ratio. The system was designed to create a monetary unit with more stability than one based on a single metal. In a bimetallic system, the ratio, which is determined by law, is expressed in terms of weight, e.g., 16 oz of silver equal 1 oz of gold, or a ratio of 16 to 1. The legal ratio has no relationship to the commercial value of the metals, which fluctuates, constantly. The discrepancy between the commercial and face values of the two metals made bimetallism too unstable for most modern nations. The system was practiced in the U.S. and other countries (except England, where gold was used) in 18th and 19th cent.

Binary. Having two parts or components. A *binary compound,* for example, is any compound of only two elements. A *binary alloy* has two component metals.

Binary Star. A system of two separate but associated stars (*components*) orbiting around a common centre of mass. Three types are observed: visual *binaries* are those where the angular separation of the components is noticeable enough for a telescope to resolve them or even for the naked eye to distinguish them. *Spectroscopic binaries* are those detected by the cyclic variations in the Doppler effect caused by the successive approach and recession of the components. *Eclipsing* or *photometric binaries* are those in which a brighter component is periodically eclipsed by a darker one. Such binaries are also called *Algol variables* after the star Algol in Perseus.

Binary Star. Pair of stars that are held together by their mutual gravitational attraction and revolve about their common center of mass. True binary stars are distinct from optical doubles—pairs of stars that lie along nearly the same line of sight from the earth but are not physically associated. A visual binary is a pair of stars that can be seen be direct telescopic observation to be a distinct pair with shared motion. A spectroscopic binary cannot be distinguished telescopically as two separate stars, but spectral lines from the pair show a periodic Doppler Effect that indicates mutual revolution. An eclipsing binary has the plane of its orbit lying in the line of sight and shows a periodic fluctuation in brightness (See variable star) as one star passes in front of the other.

Binding Energy. The energy that has to be supplied to an atomic nucleus in order to separate it completely into its constituent nucleons.

Binocular. Involving the simultaneous use of both eyes. *Binocular vision* is normal vision using two eyes—the two images from slightly different viewpoints enable the observer to gain an impression of depth. *Binoculars* (or *field glasses*) have two terrestrial telescopes mounted side by side with inverting lenses or prisms to produce an erect image. In prism binoculars the increased path length gives the instrument a longer focal length.

Binoculars. Small optical instrument, consisting of a pair of Telescopes mounted on a single, usually adjustable frame, that is used for magnifying distant objects. Light entering each telescope through its objective Lens is bent by a pair of Prisms before passing through one or more additional lenses

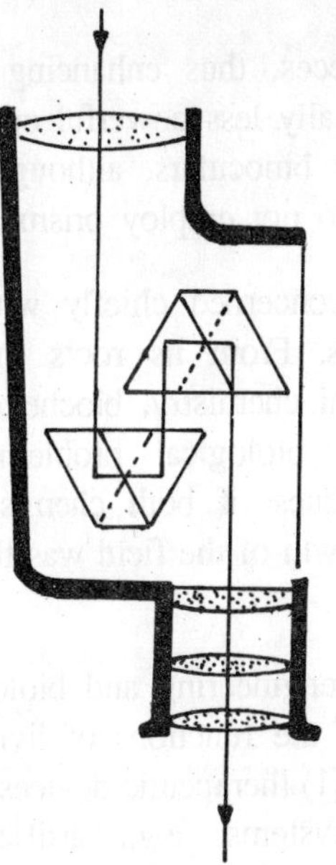

Fig. B-6. Prism binocular.

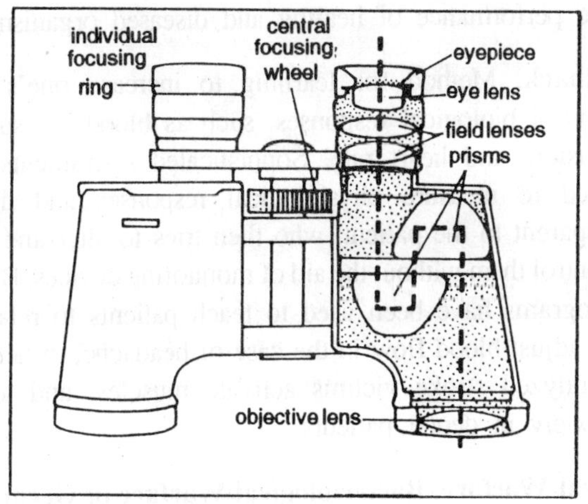

Fig. B-7. Binoculars.

in the eyepiece. The prisms turn the image—inverted by the objective lens—right side up again and allow the distance between the objective lenses to be twice as far apart as that between the eyepieces, thus enhancing the viewer's depth perception. The usually less powerful opera and field glasses are also classed as binoculars, although both use Galilean telescopes, which do not employ prisms.

Biochemistry. Science concerned chiefly with the chemistry of biological processes. From its roots in chemistry, chiefly organic and physical chemistry, biochemistry has broadened to encompass any biological problem amenable to the investigative techniques of both chemistry and physics. A milestone in the growth of the field was the elucidation of the structure of DNA.

Bioengineering. Use of engineering and biological principles for the identification of the functions of living systems and for the development of (1) therapeutic devices, especially artificial body parts and systems, e.g., artificial blood vessels, Pacemakers, Dialysis equipment, and artificial limbs that function like their prototypes; and (2) equipment for monitoring the performance of healthy and diseased organisms.

Biofeedback. Method for learning to increase one's ability to control biological responses, such as blood pressure, muscle tension, and heart rate. Sophisticated instruments are often used to measure physiological responses and make them apparent to the patient, who then tries to alter and ultimately control them without the aid of monitoring devices. Biofeedback programs have been used to teach patients to relax muscles or adjust blood flow in the case of headache, to help partially paralyzed stroke victims activate muscles, and to alleviate anxiety in dental patients.

Biological Warfare, Bacteriological Warfare or **Germ Warfare.** Employment in war of microorganisms to injure or destroy human beings, animals, or crops. Although "first use" of both biological and chemical weapons (See chemical warfare) was prohibited by the 1925 Geneva Convention, several nations subsequently conducted research into detection and defense systems, and developed microorganisms (including strains of smallpox and the plague) suitable for military retaliation. Such organisms can be delivered by animals, especially rodents or insects, or by Aerosol packages—built into artillery shells or the warheads of ground-to-ground or air-to-ground missiles and released into the atmosphere to infect by inhalation. Research in this area did not end after 1972, despite an agreement that year by more than 100 nations, including the U.S. and the USSR, to prohibit the development, testing, and stockpiling of biological weapons.

Biology. Science dealing with living things, broadly divided into Zoology, the study of animal life, and Botany, the study of plant life. Subdivisions include Anatomy and Physiology; Genetics; molecular biology, the study of cells (cytology), tissues (histology), embryonic development, and microscopic forms of life; as well as Classification, Evolution, and paleontology (the study of Fossils).

Bioluminescence. Production of light by living organisms resulting from the conversion of chemical energy to light energy. Bioluminescent plants include certain Mushrooms and Bacteria that emit light continuously. The dinoflagellates, a group of marine algae, produce light only when disturbed. Bioluminescent animals include such organisms as Comb Jellies, annelid worms, Mollusks, insects such as fireflies, and Fish. Some animals seem to use luminescence in courtship and mating, to divert predators, or to attract prey.

Bionics. Study of living systems with the intention of applying their principles to the design of engineering systems. Drawing on interdisciplinary research in the mechanical and life sciences, bionics has been used to develop audiovisual equipment based on human eye and ear function, to design air and naval craft patterned after the biological structure of birds and fish, and to incorporate principles of the human neurological system in data-processing systems.

Biophysics. Application of various tools, methods, and principles of physical science to the study of biological problems. In biophysics, physical mechanisms and mathematical and physical models have been used to explain life processes such as the transmission of nerve impulses, the muscle contraction mechanism, and the visual mechanism.

Biopsy. Examination of cells or tissues removed from a living organism to aid medical diagnosis. Samples may be removed surgically, as in excision of breast tissue, or by aspiration of cells through a special needle, as in the case of bone marrow.

Biorhythm or **Biological Rhythm.** Cyclic pattern of changes in physiology or in activity of living organisms, often synchronized with daily, monthly, or yearly environmental changes. Rhythms that vary according to the time of day (circadian rhythms), in part a response to daylight or dark, include the opening and closing of lowers and the night time increase in activity of nocturnal animals. circadian rhythms also include activities that occur often during a 24-hour period, such as blood pressure changes and urine production. Annual cycles, called cirannual rhythms, respond to changes in the relative length of periods of daylight and include such activities as migration and animal mating. Marine organisms are affected by tide cycles. Although the exact nature of the internal mechanism is not known, various external stimuli-including light, temperature, and gravity—influence the organism's internal clock; in the absence of external cues, the internal rhythms gradually drift out of phase with the environment.

Biosphere. Irregularly shaped envelope of the earth's air, water, and land, encompassing the heights and depths at which living things exist. The biosphere is a closed and self-regulating system sustained by grand-scale cycles of energy and materials.

Biprism. A prism made up of two very narrow prisms placed base to base. If positioned in front of a single light source it produces interference fringes.

Bird. Warm-blooded, egg-laying Vertebrate of the class Aves, having its body covered with Feathers and its forelimbs modified into wings. Like Mammals, birds have a four-chambered heart; they have a relatively large brain and acute hearing but little sense of smell. Believed to have evolved from Reptiles, birds are highly adapted for flight. Their feathers, though light, protect against cold and wet and have great strength. Intricate courtship displays are performed by many species during breeding season, when birdsong is most pronounced; singing ability is usually restricted to, or superior in, the male. Most birds build some kind of nest for their eggs, which vary in size, shape, color, and number according to species. The chief domestic birds are the Chicken, Duck, Goose, Turkey, and guinea fowl. Among the game birds hunted for food and sport are Grouse, Pheasant, Quail, and duck.

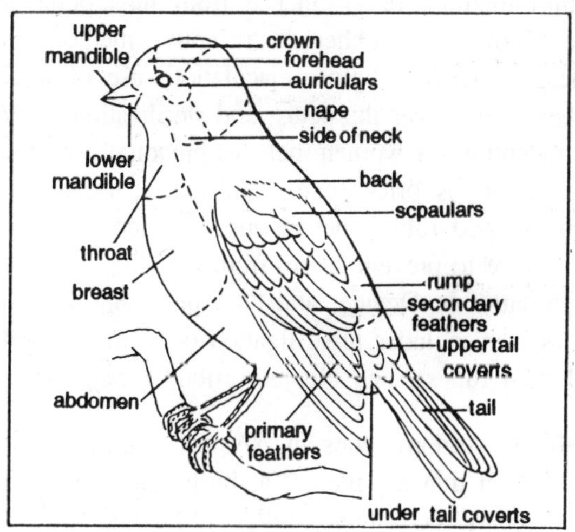

Fig. B-8. General anatomy of a bird.

Birefringence. *See* Double refraction.

Birkeland-Eyde Process. A process for the fixation of nitrogen by reacting nitrogen and oxygen in an electric arc to give nitric oxide: $N_2 + O_2 = 2NO$. [After Kristian Birkeland

(1867-1917) and Samuel Eyde (1866-1940). Norwegian scientists].

Birth Control. Practice of preventing conception for the purpose of limiting the number of births; also called contraception and family planning. The modern movement for birth control began in Britain, where the writings of Malthus stirred interest in the problems of overpopulation. The first birth-control clinic in the U.S. was opened in 1916 by birth-control leader Margaret Sanger. On an international level, birth control is led by the International Planned Parenthood Federation (founded 1952), and in some countries (e.g., Sweden, Japan, and many Communist nations) the government provides birth-control assistance to its people in order to limit population growth. The Roman Catholic Church has provided the main opposition to the birth control movement, approving only the so-called rhythm method, or abstinence from intercourse around the time of ovulation. Other contraceptive methods include, for males, withdrawal before ejaculation; use of a condom, or rubber sheath over the penis; and sterilization by Vasectomy, contraception for women includes precoital use of spermicidal vaginal suppositories, foams, and jellies; use of a diaphragm, a cup-shaped rubber device inserted into the vagina before intercourse to prevent sperm from traveling to the egg; use of an Intrauterine Device; and "the pill". Surgical sterilization for women includes tubal ligation (severing or sealing of the Fallopian tubes). See also Abortion; population.

Birth defects. Abnormalities in physical and/or mental structure of function that are present at birth. They range from minor to seriously deforming and/or life-threatening, with some major defect occurring in approximately 3% of births. Defects may be genetic in origin (as in Down's Syndrome, Tay-Sachs Disease, Sickle-Cell Anemia, and Hemophilia) or may be caused by infectious agents (e.g., Herpes Simplex, Rubella, Venereal Diseases). Other teratogenic (malformation causing) agents include drugs or hormones taken by the mother (e.g., thalidomide; Des) and maternal illnesses (e.g., diabetes). The mother's nutrition, alcohol and smoking habits, and exposure to radiation can also affect the developing fetus. Certain birth defects can now be detected prenatally through the procedure of Amniocentesis, and surgical procedures are being explored to correct certain disorders before birth.

Bisector. A line or plane that divides an angle,line, solid, etc., into two equal parts.

Bismuth. Symbol: Bi. A pinkish-white brittle metallic element, usually obtained as a by-product in refining other metal ores. It expands when solidified, making it suitable for use in alloys for casting. Its main use is in the production of low-melting alloys, such as Wood's metal. Chemically, it resembles arsenic and antimony, forming covalent bismuth III (*bismuthous*) and bismuth V (*bismuthic*) compounds. Bismuth III compounds can also be ionic, containing the Bi^{3+} ion. Bismuth salts tend to hydrolyse in solution to yield insoluble bismuthyl compounds. A.N. 83; A.W. 208.98; m.pt. 271.3°C; r.d. 9.75.

Bismuth Oxide Chloride (bismuth oxychloride). A white lustrous crystalline powder, BiOCl. It is the insoluble product of hydrolysis of bismuth (III) chloride and is used in face powder and as a pigment. r.d. 7.72.

Bison. hoofed Mammal (genus *Bison*) of the Cattle family, with short horns and heavy, humped shoulders. The European wisent *(B. bonasus)* is larger than the North American bison *(B. bison),* commonly called Buffalo, which may reach a shoulder height of 5 ft (1.5 m) and a weight of 2,500 lb (1,130 kg). Bison roamed America in vast herds until slaughter by settlers for sport and meat reduced them to near extinction; they are now protected and thriving.

Bit. A unit of information; the minimum amount of information necessary to specify either of two alternative states. In particular, a bit is a binary digit (0 or 1) used in binary notation. See *also* byte.

Bittern. The liquid remaining after sodium chloride has been crystallized from sea water. It is a source of iodine, bromine, and some magnesium salts.

Bittern. migratory marsh Bird of the Heron family. The American bittern, widely distributed in E North America, is 2 to 3 ft (61 to 91 cm) tall. When pursued, the bittern escapes detection by standing motionless with uplifted bill, its yellow and brown markings blending with the marsh grasses.

Bittersweet. Name for two unrelated fall-fruiting, woody vines. One (*Solanum dulcamara*) belongs to the Nightshade family. Its twigs and stems yield a medicinal narcotic poison similar to belladonna. The more popular bittersweet (*Celastrus scandens*) is in the staff tree family. Both grow in North America.

Bitumen. Any of various mixtures of hydrocarbons, especially semi-solid mixtures extracted from coal.

Bituminous. Containing bitumen or tar.

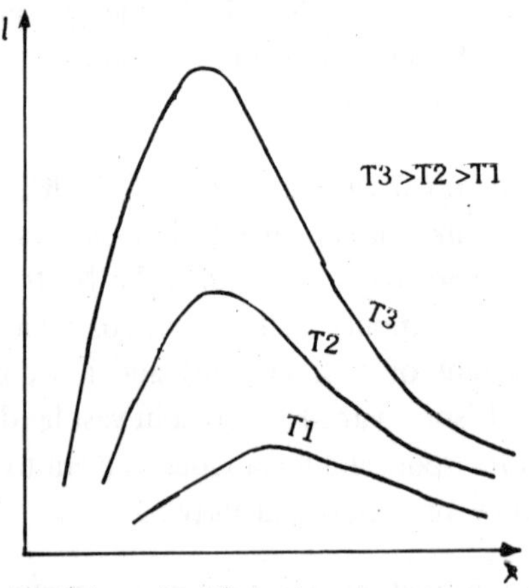

Fig. B-9. Curves of intensity (*I*) against wavelength (λ) for black-body radiation.

Bivalve. Mollusk (class Pelecypoda) with a laterally compressed body and an external shell consisting of two dorsally hinged valves. Bivalves lack eyes and tentacles; in most, a muscular foot, used for burrowing, protrudes from the valves in front of the body. Shells range in size from 1/16 in. (2 mm) to over 4 ft (120 cm). Bivalves are an important food source and include Clams, Cockles, Shipworms, Scallops, Mussels, and Oysters.

Blackbird. Perching Bird belonging to the family Icteridae, related to the Bobolink and Oriole. Except during breeding season, blackbirds travel in flocks; these may number as many as 5 million birds and often do serious crop damage. Common species are the red-winged, yellow-headed, and brewer blackbirds. The European blackbird is a Thrush.

Black body. A body that absorbs all the radiation falling on it: i.e. it has an absorptance of 1. A black body is a hypothetical ideal but a practical approximation can be made by using a small hole in the wall of a constant-temperature enclosure.

The radiation emitted from a black body has a continuous distribution of wavelengths with a characteristic intensity maximum at a particular wavelength (see illustration). The *black-body radiation* lies mainly in the infrared region at lower temperatures. As the temperature is increased the maximum in the intensity-wavelength curve moves to shorter wavelengths (*see* Wien's displacement law) until a temperature is reached at which the body is incandescent. A black body is sometimes called a *full radiator*.

Blackdamp. The air, depleted in oxygen, left in coal mines after an explosion of firedamp.

Black-eyed Susan or Yellow daisy. Weedy biennial North American daisylike wildflower (*Rudbeckia hirta*) of the Composite family, with yellow rays and a dark brown center. This and other species in the genus are also called yellow coneflowers.

Black hole. A hypothetical entity produced by the contraction of a star to such an extent that its gravitational field is high enough to prevent light from leaving it. When a star has used up its nuclear fuel it begins to contract under the influence of gravitational forces between its constituent particles. Its eventual fate depends on its mass: in particular, if its mass is large enough it undergoes a supernova explosion and the remaining core of matter forms a neutron star. This occurs if the core remaining after the supernova has a mass of about 1.4 times that of the sun. If the core has a mass several times that of the sun the gravitational forces are large enough to compress the matter still further, so that its density is even larger than that in a neutron star. Under such conditions a very large mass is concentrated into a very small volume and an immense gravitational field is produced, large enough to prevent any matter, light, or other radiation escaping. The theory of such a phenomenon is treated by the general theory of relativity but in simple terms of black hole forms when the escape velocity from a star exceeds the velocity of light. The black hole is the region in space (or more correctly in space-time) from which no matter or radiation can escape. The boundary of this region is called the *event horizon;* it is a sphere with a radius given by $2GM/c^2$, where G is the gravitational constant, M is the mass, and c is the velocity of light. This radius is called the *Schwartzschild radius* of the mass.

Although black holes are theoretical concepts there is some evidence that they exist. An isolated black hole would not be detected because it would not emit any form of radiation. However, there is some evidence for the presence of black holes in certain binary stars that emit X-rays. In the constellation of Cygnus, for example, there is a blue supergiant that revolves around a small invisible companion with a period of 5-6 days. The combination is a variable source of X-rays. The invisible companion must have a mass about 15 times that of the sun and seems to have a diameter of only about 30 miles. Its mass is too large for a stable neutron star and it is thought to be an example of a black hole, which is continually gaining matter by attraction from the supergiant. The matter would not enter the black hole directly but would form a rotating disc around it within which particles of matter would spiral in towards the black hole. In this process the matter is compressed and heated to very high temperatures, at which it emits the observed X-rays.

Within a black hole the behaviour is described by the general theory of relativity and the classical laws of physics are not obeyed. One problem concerns what happens to the collapsing matter within the hole's event horizon. In theory, it should contract indefinitely, finally disappearing at a point at which the curvature of space-time becomes infinite—such a point is called a *singularity.* The singularity that would be formed by a collapsing star is surrounded by an event horizon—a condition known an *cosmic censorship.* Some physicists have postulated the existence of *naked singularities,* in which the point of infinite curvature is not surrounded by an event horizon. Such an entity might "create" matter and light by the reverse process of that occurring in a black hole. The result would be a *white hole.* In particular, a naked singularity might have been the source of the "big bang" postulated in the big-bang theory of the origin of the universe. However, the behaviour of a system under the bizarre conditions applying at a singularity cannot be treated by present physical theories and the existence of such entities as white holes is still no more than speculation.

Blast furnace. Structure used chiefly in Smelting, i.e., extracting Metals, mainly Iron and Copper, from their Ores. The principle involved is that of the reduction of the ores by the action of Carbon Monoxide, i.e., the removal of oxygen from the metal oxide in order to obtain the metal. Pig Iron prepared in the blast furnace is converted to Steel by the Bessemer Process. Copper ore treated in a blast furnace yields a copper sulfide mixture, which is usually further refined by electrolytic methods.

Blast furnace. A furnace for the production of pig iron from oxide ore. Typically it consists of a large steel shell lined with refractory bricks. The charge, consisting of alternate layers of ore, coke, and flux, is added to the top and preheated air is blown in through inlet pipes located near the bottom of the furnace. The temperature reaches about 1300°C and the iron oxide is reduced to iron: $2Fe_2O_3 + 3C = 4Fe + 3CO_2$. The flux is often limestone, which at the temperature used decomposes to lime and carbon dioxide: $CaCO_3 = CaO + CO_2$. The lime reacts with impurities in the ore to form a molten slag of calcium silicate, aluminate, etc. Molten pig iron is tapped from the bottom of the furnace—the process is a continuous one.

Blasting gelatin. A gelatinous elastic substance made by mixing nitroglycerin with gun cotton (nitrocellulose), used as a powerful explosive.

Blazonry. Science of describing or depicting armorial bearings. Since the Middle Ages the science has grown very complicated. Its chief part is the description of the shield of a coat of arms, including the color of the field on which devices are displayed. Arms are identified by their charges; the most common of these, the ordinaries, have lines of division, heraldic animals, and flowers. Blazonry also involves descriptions of the crest above the shield and of the motto. In England blazonry is regulated by the Heralds College.

Bleaching powder (chloride of lime). A white powder made by the reaction of chlorine on slaked lime. It has the formula $CaOCl_2$ and consists of mixed hydrated calcium chloride, $CaCl_2$, calcium hydroxide, $Ca(OH)_2$ and calcium hypochlorite, $Ca(OCl)_2$. Bleaching powder acts with dilute acids to give chlorine. It is used as a bleaching agent and disinfectant.

Blende. A mineral sulphide of a metal. Zinc blende, for example, is naturally occurring zinc sulphide, ZnS.

Blight. Any sudden, severe plant disease characterized by the withering and death of the plant; also, the causative agent of such a disease. Most blights are caused by Bacteria (e.g., bean blights), Viruses, (e.g., soybean bud blight), or Fungi (e.g., chestnut blight).

Blindness. Partial or complete loss of sight. Blindness may be congenital or caused by injury or disease. Lesions of the brain Cataract, Glaucoma, or retinal detachment can result in loss of vision, as can changes in the Eye associated with disorders such as Diabetes and Hypertension. In retinitis pigmentosa, a genetically transmitted disorder, visual impairment develops in childhood and gradually progresses to complete blindness. The Braille system enables the blind to read and write. See Guide Dog; see also Birth Defects; Color Blindness.

Blind spot. *See* eye.

Blink comparator (blink microscope). A viewing apparatus that can alternately display to the user in rapid succession each of two photographic plates of the same part of the sky, so that minute differences can be observed in the brightness, position. etc., of celestial objects.

Block book. Book printed from engraved wooden blocks, one for each page. Although produced in Europe before and after the invention of Printing, block books have a richer history in China and Japan, where the large number of written characters made printing from movable type impractical.

Block copolymer. *See* polymer.

Blood. Fluid that is pumped by the Heart and circulates through the body via arteries, veins, and capillaries, carrying oxygen and nutrients to the body tissues, and carbon dioxide and wastes away from them. It is also involved in tissue repair, cell Metabolism, infection resistance, and other life-sustaining activities. Blood is made up of plasma, a colorless fluid containing red blood cells (erythrocytes), which carry on oxygen-carbon dioxide exchange, white blood cells (leukocytes), which defend the body from foreign agents; platelets that function in blood clotting; Hormones; and essential salts and Proteins. There are about 6 quarts (5.6 liters) of blood in an average-sized adult male human. Blood is classified into Blood Groups. A deficiency of red blood cells is Anemia; abnormal proliferation of leukocytes is known as Leukemia.

Blood bank. Site for collecting processing, typing, and storing whole Blood and blood products. Whole blood may be preserved up to 21 days without losing its usefulness in Blood Transfusions; an anticoagulant is added to it to prevent clotting Blood plasma, the fluid portion of the blood, may be frozen and stored indefinitely.

Blood Groups. Classifications of human Blood based on immunological properties. The most widely used blood classification system is the ABO system, described by Land-Steiner. It divides blood into A, B, AB and O groups, depending on the presence of specific chemical substances on the surface of the red blood cells (A, B, AB), or their absence (O). These substances act antigenically, i.e., to cause the formation of specific antibodies when injected into a recipient (see Immunity). Because the formation of antibodies can create a potentially dangerous condition in an individual receiving a Blood Transfusion, the blood of a donor must be compatible with that of the recipient. An other blood-group system important in blood transfusion compatibility is the RH Factor.

Bloodhound. Large Hound; shoulder height, c. 25 in (63.5 cm); weight, 80-110 lb (36.3—49.9 km). Its short, smooth coat is black and tan, red and tan, or tawny. The skin hangs in folds on the face. The oldest hound, Known in Europe over 2,000 years ago, it was brought to the U.S. in the 19th cent. It has a superb sense of smell and is used for tracking.

Blood pressure. Force exerted by blood upon the walls of the arteries. It is initiated by the pumping action of the heart, and pressure waves can be felt at the wrist and other Pulse points. Blood pressure is strongest in the aorta, where the blood leaves the Heart, and diminished progressively in the smaller vessels. Contraction of the heart (systole) produces the highest pressure, while heart relaxation (diastole) reduces the pressure to its lowest point. Pressure is measured at the brachial artery in the forearm (for consistency) in millimeters of mercury; pressures of about 120/80 (systolic/diastolic) are considered normal in young people. Conditions involving high blood pressure include Hypertension and Stroke.

Bloodstone or **Heliotrope.** Green Chalcedony spotted with red, used as a Gem. It is found in India, the U.S., Brazil, and Australia.

Blood Transfusion. Transfer of blood from the venous system of one person to that of another, or from one animal to another of the same species. Transfusions are performed to replace a large loss of blood (as in hemorrhage, Shock, and severe Burns) and as supportive treatment in certain disorders (e.g., Hemophilia). In whole blood transfusions, the Blood Groups (including the RH Factor) must be compatible with that of the recipient; if not, red blood cells will rupture and clump, a condition that can result in Jaundice, kidney damage, and death. People with blood group O are universal donors, having blood that is compatible with all other groups; those with blood group AB are universal recipients, able to accept blood from all other groups. When whole blood is not needed or is unavailable, the fluid part of the blood (plasma) may be given.

Blueberry. Widely distributed shrubs or small trees (genus *Vaccinium)* of the Heath family, usually found in acid soils. They are related to Huckleberry and Cranberry. Blueberries are a popular food and are commercially important. The most cultivated are the high-bush blueberry (*V. cerymbosum)* and the low-bush blueberry *(V. augustifolium).*

Bluebird. Migratory Bird of the Thrush family. The Eastern bluebird *(Sialia sialis),* vivid blue with a red breast, usually nests in orchards or on the edge of woodlands; it eats insects and wild fruits.

Bluefish. Stout-bodied, delicately flavored marine Fish of the family Pomatomidae, resembling the Pompano but more closely related to the sea Bass. Averaging 30 in. (75 cm) in length and 10 to 12 lb (4.5 to 5.5 kg) in weight, the bluefish is found in warm waters of the Atlantic and Indian oceans and Mediterranean Sea. It is an excellent game fish.

Blowpipe. A pipe with a narrow nozzle for directing a jet of air into a flame and producing a more intense heat.

Boa. Live-bearing constrictor Snake of the family Boidae, which also includes the Pythons, found mostly in the Americas. Boas suffocate their prey by squeezing it. Best known are the boa constrictor *(Constrictor constrictor),* which lives in terrestrial habitats from S mexico to central Argentina and averages 6 to 9 ft (1.8 to 2.7 m); and the South American anaconda, the longest (up to 25 ft/7.9 m) and thickest (3 ft / 90 cm in girth) boa.

Board of Trade unit. Symbol: BTU. A practical unit of energy equal to 1 kilowatt-hour. It was formerly used in Britain for the sale of electricity.

Bobolink. American songbird *(Dolichonyx oryzivorus)* of the family Icteridae. The spring plumage of the male is black and white; after the breeding season, it is yellowish and brown-streaked like the female. In the north, bobolinks are insectivorous, but may feed on rice crops during southern migration; once hunted because of this destructive feeding habit, they are now protected.

Bobwhite. American henlike Bird of the Pheasant family. Bobwhites feed on insects and travel in coveys, sleeping at night in a tight circle, tails to center. In spring the coveys disperse, and each male selects a nesting territory. The characteristic call of "bob-white" functions to attract a mate and warn off other males. Hunted as game birds, bobwhites are often called Quail or Partridge. The eastern bobwhite quail *(Colinus virginianus)* is c.10 in (25 cm) long; the male has mixed brown, black, and white plumage.

Bode's law. An empirical rule relating the mean distances of the planets from the sun, having the form $x = 0.4 + 0.3 \times 2^n$, where x is the distance in astronomical units and $n = -\infty, 0, 1, 2,$ etc.

The law breaks down badly for Neptune and is completely inoperative for tune and is completely inoperative for Pluto. The fact that it also allows for a planet between Mars and Jupiter may have a bearing in our understanding of the solar system. [After Johann Elert Bode (1747-1826), German astronomer.]

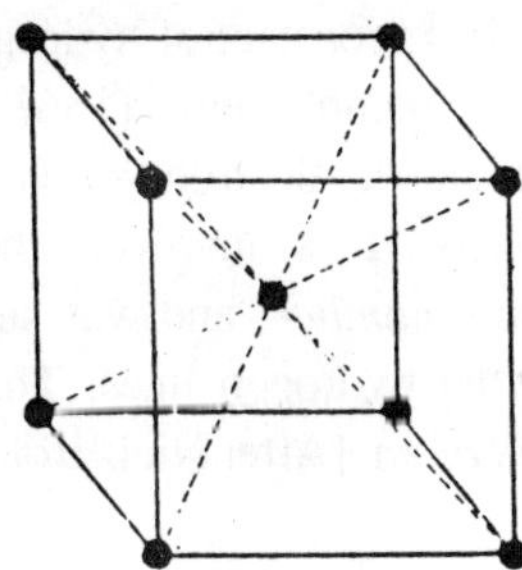

Fig. B-10. Body centred cubic unit cell.

Body-centred. Denoting a crystal structure in which the unit cell has eight atoms at its corners and one at its centre. In a *body-centred cubic* littice.

Bohr atom. A model of the atom in which the electrons move around the nucleus in circular orbits with the nucleus at the centre. The electrostatic force of attraction between the negative electrons and the positive nucleus is balanced by an outwards centrifugal force. Thus, the electrons move in orbits in an analogous way to the motion of planets in the solar system. In Bohr's theory the electrons can only move in a number of fixed orbits in which their angular momentum *(mvr)* is equal to $nh/2\pi$, where h is the Planck constant and n is an integer (1, 2, 3, ..). In these orbits the electron does not radiate electromagnetic radiation. It can be shown that the radii of the Bohr orbits are given by $r = n^2h^2\varepsilon_o/Ze^2\pi m$ where Σ_o is the electric constant and Z the atomic number. The energies are given by $E = -Ze^{\wedge} m/8n^{2'} h^2\varepsilon_o{}^2$.

The electron can occupy any of these states: normally it is in the one of lowest energy. Electromagnetic radiation is emitted or absorbed when the electron makes a transition between two orbits: the radiation is in the form of a photon of energy hv where v is the frequency. The integer n is the *principal quantum number*.

The theory gives an explanation of spectral series and a good value for the Rydberg constant of hydrogen. Later. the German physicist Arnold Sommerfeld (1868-1951) introduced a refinement in which the electrons move in ellipses precessing around the nucleus. This involved the introduction of an *azimuthal quantum number* and was successful in explaining fine structure in the hydrogen lines. This model is called the *Bohr-Sommerfeld atom.* [After Niels Bohr (1885-1962), Danish physicist.]

Bohr, Niels Henrik David (Bor), 1885-1962, Danish physicist and one of the foremost scientists in modern physics. He was professor of theoretical physics at the Univ. of Copenhagen and was later director of its Institute for Theoretical Physics, which he helped to found. Bohr was awarded the 1922. Nobel Prize in physics for his work on atomic structure. Classical theory had been able to explain the stability of the nuclear model of the Atom, but Bohr solved the problem by postulating that electrons move in restricted orbits around the atom's nucleus and explaining how the atom emits and absorbs energy. He thus combined the Quantum Theory with this concept of atomic structure.

Boiler. Device for generating steam. Two types are common: fire-tube boilers, containing long steel tubes through which the hot gases from a furnace pass and around which the water to be changed to steam circulates, and water-tube boilers, in which the conditions are reversed. A boiler must be equipped with a safety valve for venting steam if pressure becomes too great.

Boiling. The process by which a liquid changes into a gas when its vapour pressure equals the external pressure. When this occurs, small bubbles of vapour can form in the liquid and rice to the surface. The temperature at which this happens is the *boiling point* of the liquid, which depends on the forces between the molecules a substance such as water in which there are strong intermolecular forces due to hydrogen bonds has a higher boiling point than a substance such as ether, in which intermolecular forces are small. The boiling point also depends on the external pressure — the lower the pressure the lower the boiling point. Boiling points of liquids are usually quoted for standard atmospheric pressure (760 mmHg).

Boiling point. The temperature at which a substance boils, or changes from a liquid to a vapour or gas (see States of Matter), through the formation and rise to the surface of bubbles of vapor within the liquid. In a stricter sense, the boiling point of a liquid is the temperature at which its vapor pressure is equal to the local atmospheric pressure. Decreasing (or increasing) the pressure of the surrounding gases thus lowers (or raises) the boiling point of a liquid. The quantity of heat necessary to change 1 g of any substance from liquid to gas at its boiling point is known as its latent heart of vaporization.

Boiling-water reactor (BWR). *See* fission reactor.

Bolide. A large and spectacularly bright meteor that appears like a fireball and may sometimes explode with a loud detonation.

Bolometer. An instrument for measuring radiant energy, consisting of a thin strip of platinum blackened on one side and connected into a Wheatstone bridge circuit. The increase of temperature caused by radiant heat falling on the strip is measured by the change of electrical resistance.

Boltzmann, Ludwig. 1844-1906, Austrian physicist, known for his important contributions to the Kinetic-molecular Theory of Gases. By investigating the relationship between the temperature and the energy distribution of molecules in a gas, he laid the foundations of statistical mechanics. In 1883 he demonstrated theoretically a law (sometimes called the Stefan-Boltzmann law) describing the radiation from a Blackbody that had earlier been found experimentally by the Austrian physicist Josef Stefan.

Boltzmann constant. Symbol: *k*. A constant equal to the gas constant *R* divided by the Avogadro constant *L*. It is equal to 1.380 54 joules per kelvin. [After Ludwig Boltzmann (1844-1906), Austrian physicist.]

Bolyai, Janos or **Johann** (bo'lyoi), 1802-60. Hungarian mathematician. In 1823 he independently developed hyperbolic geometry, a form of Non-Euclidean Geometry. His father, *Farkas,* or *Wolfgang, Bolyai,* 1775-1856, was also a Hungarian mathematician. Farkas's *Tentamen* (1823-33), a systematic treatment of geometry, arithmetic, algebra, and analysis, contains as an appendix his son's theory of absolute space.

Bombax. Common name for the Bombacaceae, a family of tall, thick-trunked deciduous trees, found chiefly in the American tropics. Many members of the family are commercially important, e.g., the Baobab, Balsa, or corkwood, which yields the lightest lumber in the world; and the ceiba tree source of Kapok.

Bomb calorimeter. A device for determining heats of combustion, consisting of a sealed steel vessel in which the sample is ignited in an atmosphere of oxygen. Complete combustion occurs and the heat evolved can be calculated from the rise in temperature. The apparatus is extensively used for measuring the calorific value of fuels and of foods. The heat of combustion is measured at constant volume—the pressure rises during the reaction.

Bond. A linkage holding together two atoms in a molecule or holding atoms or ions in a solid. The types of chemical bond are the covalent bond, electrovalent bond, coordinate bond, metallic bond, and hydrogen bond.

Bond angle. The angle between two chemical bonds taken as the angle between lines through the nuclei of each pair of atoms.

Bond energy. The energy associated with a chemical bond between two atoms. For a diatomic molecule, the bond energy is simply the heat required to dissociate it into atoms. In a compound such as methane, in which all four C-H bonds are equivalent, the bond energy of the C-H bond is 1/4 of the total heat of atomization. In molecules containing more than one type of bond the concept of bond energy is less well defined. Thus the energies of the C-H bonds in ethane would not necessarily be the same as those in methane. By considering large numbers of similar compounds it is possible to assign average values to types of bonds so that for any molecule the sum of the bond energies approximately equals the heat of atomization. The bond energy differs from the *bond dissociation energy,* which is the energy to break a bond in a compound. Thus, in ethane the bond dissociation energy of the C-C bond is the energy required to break ethane into two methyl groups.

Bond length. The distance between the nuclei of two atoms in a compound. It depends on the type of compound. For example the C-H bond in methane is longer than that in benzene.

Bone. Hard substance that forms the Skeleton in vertebrate animals. Bone consists of a gelatinous organic material called collagen, together with minerals (mainly calcium and phosphorus). In the very young, the mineral content is low and the bones are mostly cartilage, which is pliable. With age, as their mineral content increases, bones become more brittle. Bone tissue has a three-layered structure: the spongy inner layer; the compact layer surrounding the inner layer, providing support for the body; and the tough outer membrane. The inner spaces of long bones, as those in the arms and legs, filled with marrow, important in the formation of Blood cells.

Bone Black. *See* animal charcoal.

Bonifacio, Jose. (Boneta seo), 1763-1838, Brazilian scientist and architect of Brazilian independence. An eminent geologist, he influenced the Portuguese prince regent, Dom Pedro, to declare (1822) Brazil an independent monarchy with himself as Emperor Pedro I. Bonifacio served as first minister (1822-23) of the new empire, and many of his ideas were included in the 1824 constitution.

Bonsai. The art of cultivating dwarf trees, and the plants developed by this method. Bonsai, developed over 1,000 years ago in Japan, derives from the Chinese practice of growing miniature plants. In bonsai cultivation, the plants are kept small and in true proportion to their natural models by growing them in small containers, feeding and watering only enough for healthy growth, pruning, and forming branches to the desired shape by applying wire coils. The selection of containers, the plant's position in a container, and the choice of single plants or a group are important aesthetic considerations.

Booby. Large, streamlined sea Bird of the family Sulidae, called gannets in northern waters. Boobies have heavy bodies, long, pointed wings, long, wedge-shaped tails, and short, stout legs. They fish by diving on their prey from great heights and pursuing it under water. Gullible and unwary, as its name indicates, the booby is easy prey for hunters and is diminishing in number.

Boole, George, 1815-64, English mathematician and logician. Boolean algebra, his form of symbolic Logic, is central to the study of the foundations of pure mathematics and is the basis of computer technology. Boole wrote *An Investigation of the Laws of Thought* (1854), as well as works on calculus and differential equations.

Boolean algebra. An abstract mathematical system primarily used in Computer science and in expressing the relationships between Sets (groups of objects or concepts). The notational system was developed by the English mathematician George Boole about 1850 to permit an algebraic manipulation of logical statements. Such manipulation can demonstrate whether or not a statement is true and show how a complicated statement can be rephrased in a simpler, more convenient form without changing its meaning. When used in set theory, Boolean algebra can demonstrate the relationship between groups, indicating what is in each set alone, what is jointly contained in both, and what is contained in neither. The expression of electrical networks in Boolean notation has aided the development of switching theory and the design of computers.

Boracic acid. *See* boric acid.

Borate. Any salt or ester of boric acid.

Borage. Common name for the Boraginaceae, a family of widely distributed herbs and tropical shrubs and trees characterized by rough or hairy stems; four-part fruits; and, usually, fragrant blossoms. Species of forget-me-not *(Myosotis)* and species of heliotrope *Heliotropium)* are cultivated in and native to North America.

Borax or sodium tetraborate decahydrate. Chemical compound ($Na_2B_4O_7$. $10H_2O$) occurring as a colorless, crystalline salt or a white powder. Borax is used as an antiseptic, cleansing agent, water softener, corrosion inhibitor in antifreeze, and flux for silver soldering, and in the manufacture of fertilizers, Pyrex glass, and pharmaceuticals.

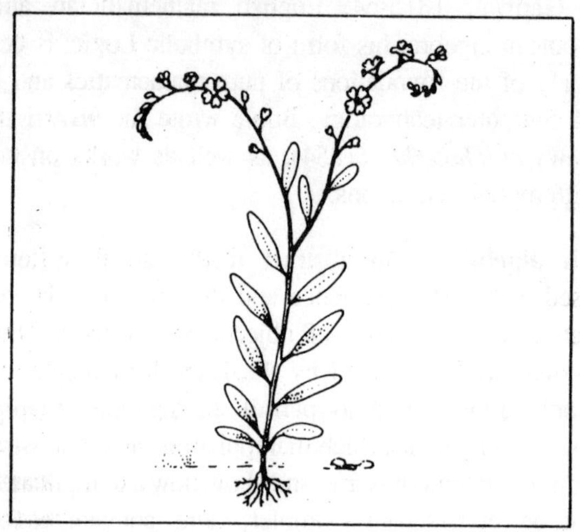

Fig. B-11. *Borage: Forget-me-not,* Myosotis virginica, *a plant of the borage family.*

Borax (disodium tetraborate). A whitish crystalline hydrated solid, $Na_2B_4O_7$.10-H_2O, found in some salt lakes and alkaline soils. The pentahydrate and the anhydrous salt can also be prepared. Borax is used in the manufacture of glass and enamels. M.pt. 741° C (anhydrous); r.d. 1.7 (decahydrate).

Borax-bead test. A test for the presence of some metals in salts. The sample is mixed with borax and fused in a flame on the end of a piece of wire. A small bead forms with a colour that may be characteristic of the metal present. Cobalt salts, for example, form a blue bead.

Borazon. *See* boron nitride.

Bordeaux mixture. A liquid mixture of slaked lime and copper sulphate solution used as a fungicide and insecticide for plants.

Borden, Gail. 1801-74, American inventor; b. Norwich, N.Y. His process (patented 1856) of evaporating Milk was found to be of great value for the army during the Civil War, and its use spread rapidly afterwards. Borden subsequently also patented processes for concentrating fruit juices and other beverages.

Boric acid (orthoboric acid, horacic acid). A colourless odourless crystalline solid or powder, H_3BO_3. It is a very weak acid prepared by the addition of hydrochloric acid to borax and is used in ceramics and fire-proofing compositions. M.pt. about 160° C; r.d. 1.44.

Boride. Any compound of boron with a more electropositive element. Most metals form borides. They are chiefly hard interstitial materials with metallic conductivity.

Born, Max. 1882-1970, German physicist. For his statistical interpretation of quantum mechanics (see Quantum theory), he shared the 1954 Nobel Prize in physics. Born was head (1921-33) of the physics department at the Univ. of Gottingen. After Nazi policies forced him to leave Germany, he taught at Cambridge Univ. and the Univ. of Edinburgh before returning to Germany in 1954.

Boron. (B), Nonmetallic element, isolated by Sir Humphry Davy in 1807. As a dark-brown to black amorphous powder, boron is more reactive than its jet-black to silver-gray crystalline metallike form. Borax and boric acid are common compounds. Boron is used in the shielding material and in some control rods of nuclear reactors.

Boron. Symbol: B. A nonmetallic element found naturally in several minerals, especially borax ($Na_2B_4O_7 . 10H_2O$). Usually, it is obtained as a yellow brown amorphous powder but a yellow crystalline form also exists. It can be made by reducing the oxide with magnesium metal and is used in hardening steels, in control rods for nuclear reactors, and in semiconductors. Boron is a fairly unreactive element. Its most important compounds are boric acid and borax. Many metal borides exist, most of them prepared by direct combination of the elements at high temperatures. Boron forms covalent compounds with a valency of three, although the boron hydrides are electron deficient. A.N. 5; A.W. 10.811; m.pt. 2300°C; sublimes at 2550° C; r.d. 2:34.

Boron nitride. A crystalline solid, BN, made by direct reaction of the elements at high temperatures and pressures. It has two crystalline structures, one resembling that of diamond and the other that of graphite. The former dimorph, *borazon,* is one of the hardest substances known. M.pt. 3000° C; r.d. 3.48 (borazon).

Boron trifluoride. A colourlsss gas, BF_3, with a pungent suffocating odour, used as a catalyst in Friedel-Crafts reactions. Boron trifluoride can act as a Lewis acid. M.pt. - 127°C; b. pt. - 100°C.

Borzoi. Tall, swift Hound; shoulder height, 26-31 in (66-81.2 cm); weight, c. 85 lb (38.6 kg). Its long, silky coat may be flat, wavy, or curly, and any color (usually white with markings). Developed in Russia in the early 17th cent. To hunt wolves and hares, it is also called a Russian wolfhound.

Bosch process. A process for the manufacture of hydrogen by reacting carbon monoxide with steam over a hot catalyst: $CO + H_2O = CO_2 + H_2$. The raw material used is water gas, a mixture of carbon monoxide and hydrogen. [After Carl Bosch (1874-1940), German chemist.]

Bose-Einstein statistics. A type of statistical mechanics applied to the distribution of particles amongst energy levels when the particles are indistinguishable from one another and any number of particles can occupy each energy state. *See also* boson. [After Satyendra Nath Bose (1894-1974), Indian physicist, and Albert Einstein.]

Bose, Sir Jagadis Chunder. 1858-1937, Indian physicist and plant physiologist. Professor of physical science (1885-1915) at Presidency College, Calcutta, he is noted for his researches in plant life, especially his comparison of the responses of plant and animal tissue to various stimuli. One of his inventions is the crescograph, a device for measuring plant growth.

Boson. Any particle that obeys Bose Einstein satistics. Examples are deuterons and photons. *Compare* fermion.

Botanical garden. Public place where plants are grown both for display and for scientific study. An *arboretum* is a botanical garden devoted chiefly to woody plants, which are often arranged in cultural or habitat groups such as rock gardens, desert gardens, and tropical gardens. Botanical gardens collect and cultivate plants from all over the world, conduct experiments in plant breeding and hybridization (see Hybrid), and maintain libraries and herbariums. The Missouri Botanical Garden (St. Louis), founded c. 1860, was the first botanical garden in the U.S.

Botany. Science devoted to the study of plants, a major branch of Biology. In the 17th cent. The work of Linnaeus on the Classification of organisms contributed greatly to the growth of the science, and the indroduction of the Microscope marked the beginning of the study of plant anatomy and cells. Modern botany has expanded into all areas of biology, e.g., plant breeding and Genetics. Practical areas of botanical study include Agriculture, Agronomy. Forestry, and Horticulture.

Botulism. Acute, often fatal food poisoning from ingestion of food containing toxins produced by *Clostridium botulinum* bacteria. Most cases are caused by canned food that has been improperly processed. The disease causes disturbances in vision, speech, and swallowing and, ultimately, paralysis of respiratory muscles, leading to suffocation. Treatment involves the administration of antitoxin as soon as possible after exposure to contaminated food.

Box. Common name for the Buxaceae, a family of trees and shrubs with leathery green leaves, native to tropical and subtropical regions of the Old World and Central America. Boxes (genus *Buxus)* are widely cultivated as hedge plants and for their closer grained, strong hardwood. Boxwood takes a high polish and is used for wood engraving, carving, and turning, and for making musical instruments.

Boyd Orr, John Boyd Orr. 1st **Baron.** 1880-1971, British nutritionist and agronomist. He made notable contributions to the science of nutrition and to the solution of world food problems, winning the 1949 Nobel Peace Prize for advocating a world food policy based on need rather than trade interests. He was director general (1946-47) of the UN Food and Agriculture Organization.

Boyle, Robert. 1627-91, Anglo-Irish physicist and chemist. Often referred to as the father of modern chemistry, he separated chemistry from alchemy and gave the first precise definitions of a chemical element, a chemical reaction, and chemical analysis. He invented a vacuum pump and used it in the discovery (1662 of what is known as Boyle's law (see Gas Laws). His diverse experimental and theoretical work supplemented Sir Isaac Newton's achievements in establishing the dominance of mechanistic theory.

Brackett series. A spectral series in the infrared emission of hydrogen with wavelengths given by $1/\lambda = R(1/4^2 - 1/m^2)$, where R is the Rydberg constant and m is 5, 6, 7, etc.

Bradley, James. 1693-1762, English astronomer. He discovered the Aberration of Starlight (announced in 1729) and the Nutation, or nodding," of the earth's axis (announced in

1748). Bradley became astronomer royal and director of the Royal Greenwich Observatory in 1742.

Bragg's law. An equation for the diffraction of X-rays by crystal planes giving the condition for maxima in the diffraction pattern; $n\lambda - 2d \sin \theta$. where θ is the wavelength, d is the distance between planes, and θ is the angle between the plane and the incident beam. The equation is derived by considering that reflection occurs from atomic planes and interference take place between X-rays reflected from different planes. *See also* X-ray crystallography, X-ray diffraction. [After Sir Lawrence Bragg (1890-1971), English physicist.]

Bragg, Sir William Henry. 1862-1942, English physicist. He was on the faculties of the Univ. of Adelaide, Australia, the Univ. of Leeds and the Univ. of London, and director from 1923 of the Royal Institution's research laboratory. With his son, Sir William Lawrence Bragg, 1890-1971, he shared the 1915 Nobel Prize in physics for studies, with the X-ray spectrometer, of X-ray spectra and of crystal structure. The younger Bragg was professor of physics at Victoria Univ. (Manchester) and Cambridge, and was director (1938-53) of the Cavendish Laboratory at Cambridge.

Brahe, Tycho. (Bra), 1546-1601, Danish astronomer. His exact observations of the planets were the basis for Kepler's Laws of planetary motion. Studies of the moon's motion and of a supernova (1572) and improvements of instruments were among his contributions. Brahe never fully accepted the Copernican System, compromising between that and the Ptolemaic System. In his system, the earth was the immobile body around which the sun revolved, and the five planets then known revolved around the sun.

Braille, Louis. 1809?-1852, French inventor of the Braille system of printing and writing for the blind. Blind from an accident at age three, he attended and later taught at the Institution des Jeunes Aveugles, Paris. He evolved a system of writing with points based on Charles Barbier's method, though much simpler. The Braille system consists of six raised points used in 63 possible combinations; it is in use, in modified form, for printing, writing, and musical notation.

Brain. Supervisory center of the Nervous System in all vertebrates. The brain controls both conscious behaviour (e.g., walking and thinking) and most involuntary behaviour (e.g., heartbeat and breathing). In higher animals, it is also the site of emotions, memory, self-awareness, and thought. It functions by receiving information via nerve cells (neurons) from every part of the body, evaluating the data, and then sending directives to muscles and glands or simply storing the information. Information, in the form of electrical and chemical signals, moves through complex brain circuits, which are networks of the billions of nerve cells in the nervous system. A single neuron may receive information from as many as 1,000 other neurons. Anatomically, the brain occupies the skull cavity (cranium) and is enveloped by three protective membranes (meninges). The adult brain weighs $2^1/_4$ to $3^1/_4$ lb (1-1.5 kg). It has several parts, each with a loosely associated function. The *brainstem* (hindbrain), monitoring involuntary activity (e.g., breathing), and the *cerebellum,* coordinating muscular movements and posture, are together the basic machinery for survival and reproduction. The *forebrain,* composed of the limbic system and cerebral cortex, regulates higher functions. The limbic system (including the thalamus, hypothalamus, pituitary, amygdala and hippocampus, and olfactory cortex) is associated with vivid emotions, memory, sexuality, and smell. The forebrain's cerebral cortex, in the uppermost portion of the skull, has some areas concerned with muscle control and the senses and others concerned with language and anticipation of action. The cerebral cortex is split into two hemispheres, each controlling the side of the body opposite to it. In addition, the right hemisphere is associated with perception of melody, nonverbal visual patterns, and emotion, while the left hemisphere is associated with verbal skills. Brain function is monitored by Electroencephalography.

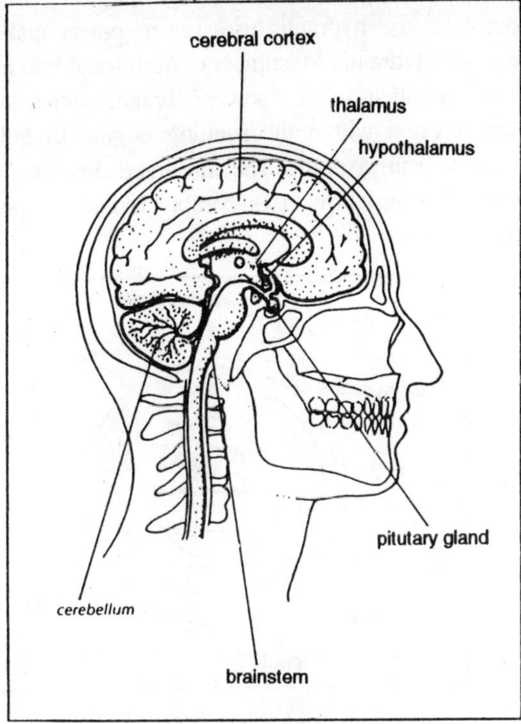

Fig. B-12. Brain.

Brake. Device used to slow or stop the motion of a mechanism or vehicle. Friction brakes, the most common kind, operate on the principle that friction can be used to convert the mechanical energy of a moving object into heat energy, which is absorbed by the brake. Friction brakes consist of a rotating part—such as a wheel, axle, disk, or brake drum—and a stationary part that is pressed against the rotating part to slow it or stop it. The stationary part usually has a lining, called a brake lining, that can generate a great amount of friction yet give long wear. The simplest brake form is the single-block brake, a wooden block shaped to fit against the rim of a wheel or drum. In disk brakes, two blocks press against either side of a disk that rotates with the wheel. Drum brakes have two semicircular brake shoe inside a rotating brake drum; when actuated, they press against the inner wall of the drum.

Automobiles use hydraulic pressure to power disk and drum brakes (see Hydraulic Machinery). Additional braking pressure may be supplied by a "power" brake, which utilizes the vacuum created within the running engine to hold a brake shoe away from a drum. The air brake, invented (1868) by George Westingshouse, uses compressed air to power block brakes on trains.

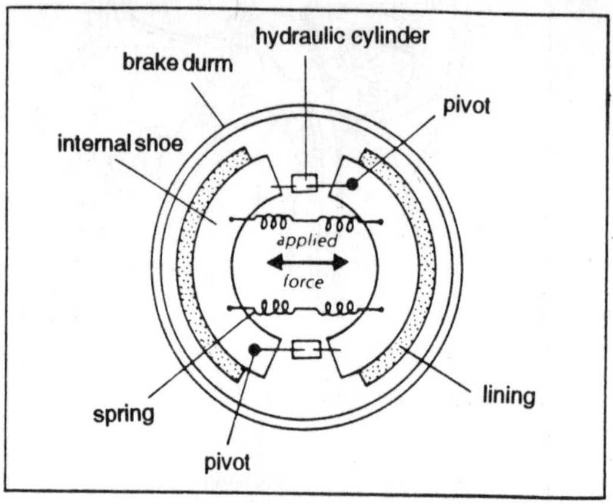

Fig. B-13. Shoe Brake.

Bramble. Plants (genus *Robus)* of the Rose family, with representatives worldwide. Members include the balckberries, raspberries, loganberries, boysenberries, and dewberries. The plants are typically shrubs with prickly stems (canes) and edible fruits that botanically are not berries but aggregates of drupelets (see Fruit). The underground parts of brambles are perennial and the canes biennial; only second year canes bear flowers and fruits. Berries are grown commercially for sale as fresh, frozen, and canned fruit, and for use in preserves, beverages, and liqueurs. Other thorny shrubs are also called brambles.

Bran. Outer coat of cereal Grains such as wheat, rye, and corn. Various brans are used as food and livestock feed, and are important in dyeing and calico printing.

Branched chain. A chain of atoms, usually carbon atoms, in a molecule, with side chains attached. Isobutane, for example, is a branched-chain hydrocarbon whereas butane has a straight chain.

Brandy. Strong alcoholic spirit distilled from wine. Manufactured in many countries, brandy is not notable in the form of cognac, made from white grapes in the Charente district of France. Most fine brandies are distilled in pot stills, blended and flavored, and stored in casks (preferably oak) to mellow. Brandies are also made from fruits other than the grape, such as plum (slivovitz) or peach.

Brass. Alloy having Copper (55% to 90%) and Zinc (10% to 45%) as its essential components. Its properties vary with the proportion of copper and zinc and with the addition of small amounts of other elements. Cartridge brass is used for cartridge cases, plumbing and lighting fixtures, rivets, screws, and springs. Aluminium brass has greater resistance to corrosion than ordinary brass. Brass containing tin (naval brass) resists seawater corrosion. Brass can be forged or hammered into various shapes, rolled into thin sheets, drawn into wires, and machined and cast.

Brass. Any of a large number of alloys containing copper and zinc (up to 40%), sometimes with small amounts of aluminium, tin, manganese, or other metals.

Brazil nut. Common name for the Lecythidaceae, a family of tropical trees. Members include the West Indian anchovy pear *(Grias cauliflora)*; several lumber trees of South America, e.g., the cannonball; and the Brazil nut trees (genus *Bertholletia)*. Brazil nut trees are found chiefly along the Amazon and Orinoco rivers. The edible nuts are oil-rich seeds that grow clumped together in hard, grapefruit-sized, woody seed pods.

Brazing. *See* solder.

Breeding. The deliberate selection of certain parent plants and animals for propagation in order to adapt them to human needs. Plants and animals have been bred selectively since their domestication in the Neolithic period. Among plants, pure lines are established by self-pollination, planting the resultant seed, and repeating the process through several generations. Crosses of pure lines produce hybrids of great vigor and uniformity. Pure-bred animals show a breeding uniformity approaching that of pure-line plants. New varieties are developed by hybridization and by breeding individuals that have developed Mutations.

Bremsstrahlung. Electromagnetic radiation produced by the deceleration of charged particles. An electron approaching an atomic nucleus loses energy and the energy lost is taken up by an emitted photon. The effect is observed for electrons of high energy. Bremsstrahlung, in the form of X-rays having a continuous distribution of wavelengths, is obtained in X-ray tubes.

Brewster's law. The principle that light reflected from a solid surface is plane polarized with a maximum polarization occurring when the angle of incidence *(a)* is $\tan^{-1}$ *(n)*, where *n* is the refractive index (i.e. $n = \tan \alpha$). The angle α is the *Brewster angle.* It is the angle at which the reflected ray makes an angle of 90° with the refracted ray. [After Sir David Brewster (1781-1868), English physicist].

Brick. Building material made by shaping clay into blocks and then hardening them in a Kiln. Sun-dried bricks are among the most ancient building materials. Examples from c.5,000 years ago have been discovered in the Tigris-Euphrates basin. The Romans faced brick buildings with stone or marble, but Byzantine and later European builders used the brick itself to provide a decorative surface. The many varieties of modern brick include firebrick, made of special clays that can be fired at very high temperatures.

Bridge. Structure spanning and giving passage over a waterway, a depression, or some other obstacle. Early bridges ranged from suspended rope walkways to arched structures of stone or brick. The mid-19th-cent. development of steel allowed for longer spans, the use of the steel truss in place of stone or masonry arches, and steel-wire suspension Cables of great tensile strength. Movable bridges, whose center portions can be raised, are generally constructed over waterways where it is impossible to build a fixed bridge high enough for water traffic to pass under it. Military bridges—temporary structures that can be erected rapidly—include the pontoon bridge, which uses air-filled floats called pontoons to support the roadway.

Bridge. Any network of four resistors, inductors, capacitors, or other circuit elements in the form of a quadrilateral with an input connected across two opposite corners and the output across the other two corners. Bridges are often used as measuring circuits, as in the Wheatstone bridge.

Brinell test. A test determining the hardness of metals by forcing a small hard steel ball into the surface. The hardness is characterized by the *Brinell number*— the ratio of the load in kilograms to the area of the depression made (After Johann August Brinell (1849-1925), Swedish engineer].

Britannia metal. An alloy of tin (80-90%), antimony (5-15%), and copper (0—3%), used in bearings.

British thermal unit. Symbol: Btu. A unit of energy equal to 1055.055 853 joules. Formerly, the British thermal unit was defined as the amount of heat required to raise the temperature of one pound of water through one degree Fahrenheit.

Brittany spaniel. Medium-sized Sporting Dog; shoulder height, c. 19 in. (48.3 cm); weight, 30-40 lb (13.6-18.1 kg). Its dense, flat or wavy coat is dark orange and white or liver and white. It dates back hundreds of years, to France and Spain, and is the only spaniel that points its quarry.

Brittle star. Marine invertebrate animal (class Ophiuroidea). An Echinoderm, it has five long, slender arms that radiate from a central disk, and the water-vascular system and tube feet common to echinoderms. Their arms break off readily, but new arms are easily regenerated. Most brittle stars are less than 1 in. (2.5 cm) across the central disk.

Broccoli. Variety of Cabbage grown for the edible immature flower panicles. It is the same variety as cauliflower *(Brassica oleracea botrytis)* and similarly cultivated.

Broglie, Louis Victor, prince de. (brogle), 1892-, French physicist. From his hypothesis that particles should exhibit certain wavelike properties, wave mechanics, a form of quantum mechanics, was developed (see Quantum Theory). Experiments proved (1927) the existence of these waves, and he won the 1929 Nobel Prize in physics for his theory.

Bromate. Any salt of bromic acid.

Bromeliad. Common name for some members of the family Bromeliaceae, chiefly epiphytic herbs and small shrubs native to the American tropics and subtropics. A typical bromeliad has strap-shaped leaves clustered in a rosette around a central cup; the flowers are found in the central cup or above it on a spike. The central cup retains water, enabling tropical bromeliads to survive dry spells. Bromeliad species such as those of the genus *Aechmea* and *Cryptanthus* are often grown as house and conservatory plants for their colorful flowers and foliage. Other members of the family include Spanish Moss and the Pineapple.

Bromeosin. *See* eosin.

Bromic acid. A colourless unstable acid $HBrO_3$, made in dilute solution by the action of sulphuric acid on a bromate salt. It is an oxidizing agent.

Bromide. Any of certain compounds containing bromine. *See* halide.

Bromination. A chemical reaction in which one or more bromine atoms are introduced into an organic compound. The term is usually used to describe reactions involving elemental bromine. An example is the bromination of ethane, under the influence of light, to give bromoethane: $C_2H_6 + Br_2 + C_2H_5Br + HBr$.

Bromine. (Br), volatile liquid element, discovered in 1826 by Antoine J. Balard. A Halogen, it is a reddish-brown fuming liquid with an offensive odor. Bromine corrodes the skin, and its vapor irritates the eyes and the membranes of the nose and throat. The only nonmetallic element that is liquid under ordinary conditions, bromine occurs in compounds in seawater, mineral springs, and salt deposits. Its compounds are used in photographic film, in flame retardants, and in conjunction with an anti-knock compound in gasoline.

Bromine. Symbol: Br. A heavy volatile reddish-brown liquid element producing a pungent irritating red vapour. It occurs in natural bromide salts and is manufactured from sea water by displacement with chlorine. Bromine is used in the manufacture of organic chemicals. It is a reactive element belonging to the halogen group. A.N. 35; A.W. 79.904; m.pt. -7.2°C; b.pt.. 58.78°C; r.d. 3.12 (liquid); valency 1, 3, 5, or 7.

Bromoethane (ethyl bromide). A colourless flammable liquid alkyl halide, C_2H_5Br, prepared from ethene and hydrogen bromide. It is used in organic synthesis and as a refrigerant. M.pt. -119°C; b.pt. 38.4°C; r.d. 1.43.

Bromoform (tribromomethane). A colourless liquid, $CHBr_3$. M.pt. 8°C; b.pt. 150°C; r.d. 2.9.

Bromomethane (methyl bromide). A colourless volatile nonflammable liquid alkyl halide,, Ch_3Br, used as a fumigating agent and solvent. M.pt. -93°C; b.pt. 4.5°C; r.d. 1.73.

Bronchitis. Inflammation of the bronchial tubes caused by viral or bacterial infection or by the inhalation of irritating fumes (e.g., tobacco smoke, air pollutants). Symptoms include cough, fever, and chest pains. Acute bronchitis may subside, or, particularly with continued exposure to irritants, may persist and progress to chronic bronchitis or Pneumonia, Bronchitis can be treated with antihistamines, cough suppressants, bronchodilators, or antibiotics.

Bronze. Any of various alloys of copper. The term was originally used for alloys of copper and tin (up to 30%), sometimes with smaller amounts of other elements. *Phosphor bronze,* for example contains small amounts (up to 1%) of phosphorus. The term *bronze* has been extended to certain copper alloys that contain no tin, as in *aluminium bronze.*

Bronze Age. Technological period when metals were first used to make tools and weapons. The earliest stage, when pure copper and bronze were used interchangeably, has been called the Copper Age. Casting was well established in the Middle East by 3500 B.C.; in the New World the first bronze was cast A.D. c. 1100. Development of a metallurgical industry coincided with urbanization, support of an artisan class, and trade for raw materials, laying the foundation for the Iron Age.

Bronze sculpture. Bronze is an ideal alloy for casting, engraving, and repousse work. When melted, it flows into the crevices of a mold, reproducing every detail; once hardened, it is easily worked with a tool. The ancient Egyptians used bronze for utensils, armor, and statuary, and outstanding Oriental bronzes have been made since ancient times. Unexcelled in bronze sculpture, the Greeks created such masterpieces as *The Zeus of Artemesium* (National Mus., Athens) and *The Delphic Charioteer* (Delphi Mus.). Copying Greek models, the Romans made thousands of bronze sculptures, as well as utilitarian objects. Most medieval bronzes were utensils and ornaments. During the Renaissance, Italian sculptors wrought magnificent bronzes, such as the Ghiberti doors to the baptistery of Florence and sculpture by Donatello, Verocchio, and Cellini. In the 18th cent. France was well known for gilded bronze furniture mounts. Major modern sculptors who have worked in bronze include Rodin, Epstein, Brancusi, Henry Moore, and Lipschitz.

Brownian movement. Small random motions of colloidal particles suspended in a liquid or gas. It is caused by random bombardment of the particles by molecules of the liquid or gas. [After Robert Brown (1773-1858), Scottish botanist.]

Brown, Robert. 1773-1858, Scottish botanist and botanical explorer. He went as a naturalist and collector to Australia (1801) and described its flora in his *Prodromus florae Novae Hollandiae* (1810). Librarian to the Linnaean Society and British Museum Curator, he observed Brownian Movement in 1827 and discovered the cell nucleus in 1831.

Brunswick green. A green pigment made by mixing chrome yellow and Prussian blue. It is used in paints.

Brush. An electrical contact on an electric motor or generator.

Brush discharge. A type of gas discharge occurring near sharp points at high voltages, the voltage being lower than that producing a spark or arc. The discharge is seen as short luminous streamers at the tip of the point.

Bubble Chamber. A device for detecting and investigating elementary particles by the tracks they make in a liquid. Usually liquid hydrogen is used, kept under pressure at a temperature slightly above its boiling point. The operating technique is to reduce the pressure for a few milliseconds, during which a photograph is taken, and reimpose the pressure before the liquid boils. Particles passing through the chamber when the photograph is taken cause a track of small bubbles along their path. Decays and interactions of particles can be investigated. Usually large numbers of photographs are taken from different positions and known electric and magnetic fields are applied, so that the particles can be identified from the curvature of their tracks.

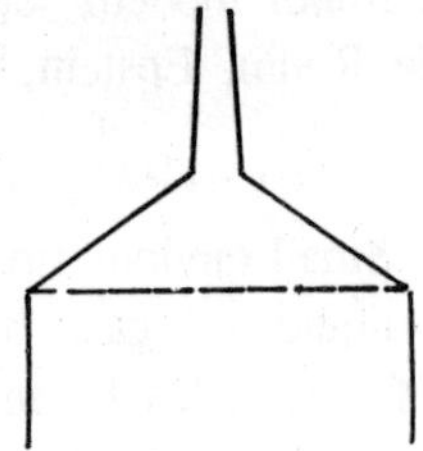

Fig. B-14. Buchner funnel.

Buchner funnel. A type of funnel with a flat perforated tray on which is placed a circular piece of filter paper. Buchner funnels are usually made of porcelain. They are used for filtering liquids by suction. [After Eduard Buchner (1860-1917), German chemist.]

Buckhorn. Common name for some members of the Rhamnaceae, a family of widely distributed woody shrubs, small trees, and climbing vines. The buckthorns (some species of the genus *Rhamnus*) and the jujube (*Zizyphus jojuba*) are cultivated as ornamentals. Jujube was also used as a flavouring and confectionery. Some members of the family yield dyes; others are used for lumber, e.g., cogwood. Other American species of *Rhamnus* are the redberry, the Indian cherry, and *R. purshiana,* which yields the purgative cascara sagrada.

Buckwheat. Common name for some members of the Polygonaceae, a family of herbs and shrubs found chiefly in north temperate areas and having a characteristic pungent juice containing oxalic acid. Members of the family include the knotweeds and smartweeds (genus *polygonum* or *persicaria*), some sorrels (the common name used also for the unrelated Oxalis), and the economically important rhubarb (genus *Rheum)* and buckwheat *genus Fagopyrum).*

Buffalo. Name commonly applied to the North American Bison but correctly restricted to related African and Asian hoofed Mammals of the Cattle family. The water, or Indian, buffalo *(Bubalus bugalis)* is a large, strong, dark gray animal with widespread curved horns, domesticated for draft. Wild forms live near rivers, where they wallow; they feed on grass and have fierce tempers.

Buffer. Solution that can keep its pH, i.e., its relative acidity of alkalinity, constant despite the addition of strong acids or bases (see Acids and Bases). Buffer solutions contain either a weak acid or weak base and one of their salts. See article on pH.

Buffer solution. A solution with a definite pH that does not change on dilution or addition of (small) amounts of acid or alkali. Buffer solutions can be made by mixing a solution of a salt of a weak acid or base with the free acid or base in calculated amounts to produce the desired pH. an example of a buffer solution is one containing acetic acid and sodium acetate. The acid partially dissociates to produce hydrogen ions: $CH_3COOH = CH_3COO^- + H^+$. The sodium acetate is a source of more acetate ions. If acid is added, the hydrogen ions combine with the acetate ions to give undissociated acetic acid. The presence of the sodium acetate "buffers the solution against increase in acidity. Buffer solutions are useful in experiments in which a contant pH is required. Buffers are also important in biological systems, which often only function under neutral on nearly neutral conditions.

Bug. Name correctly applied to Insect of the order Hemiptera, although other insects are also referred to as bugs. True bugs have partially thickened, partially membranous front wings and piercing-sucking mouthparts in the form of a beak. Most bugs and terrestrial but many are aquatic (e.g., Water Bugs). Most suck plant juices (e.g., squash bug); some suck the blood of other insects and spiders. Others feed of humans and other animals (e.g., Bedbug).

Bulb. Thickened, fleshy plant bud, usually formed below the soil surface, which stores food from one blooming season to the next. Bulbs have either layers (e.g., onion and hyacinth) or scales (e.g., some lilies). Structures that serve similar functions include the Tumber of the potato and the Rhizomes of some irises. All are specialized Stems.

Bulk modulus. *See* modulus.

Bull. Letter containing an important pronouncement of the pope. In modern times encyclicals (letters to all bishops) are usually used for doctrinal statements, whereas bulbs are employed for solemn or grave pronouncements.

Bulldog. Thick-set Nonsporting Dog; shoulder height, 13-15 in (33-38.1 cm); weight, 40-50 lb (18.1-22.7 kg). Its short flat-lying coat is brindle, white, red, or fawn. It has a low-slung body and undershot jaw, and was developed in Britain for bullbaiting and pit-fighting.

Bull terrier. Large muscular Terrier; shoulder height, 19-22 in (48.3-55.9 cm); weight, 30-36 lb (13.6-16.3kg). Its short, flat-lying, harsh coat is glossy white or brindle with white markings. It was developed for dogfighting in England c.1835.

Bumping. Violent boiling of a liquid caused when the liquid has been heated to a temperature above its normal boiling point, so that when the bubbles do form they have a higher pressure than atmospheric pressure.

Bunsen burner. A gas burner having a vertical metal tube with an air inlet at the bottom. The mixture of gas and air gives a flame with two regions. The cooler inner cone is pale blue and contains incompletely burnt gas, producing a reducing atmosphere. The hotter outer flame is almost invisible. Here all the gas is fully oxidized. The tip of this outer flame is the fully oxidized. The tip of this outer flame is the hottest part, at a temperature of about 1500°C. [After Robert Bunsen (1811-99), German chemist.]

Bunsen. Robert Wilhelm. 1811-99, German scientist. A professor (1852-89) at Heidelberg, he studied organic compounds of arsenic and developed a method of gas analysis from studies on blast furnaces. With Gustav Kirchhoff he discovered by spectroscopy the elements Cesium and Rubidium. He invented and improved various kinds of laboratory equipment, e.g., the Bunsen burner.

Buoyancy. The upthrust on a body that is partially or totally immersed in a fluid.

Burette. A long graduated glass tube with a tap at one end used to deliver known volumes of liquid in titrations, etc.

Burn. Injury resulting from exposure to heat, electricity, radiation, or caustic chemicals. First-degree burns, characterized by simple reddening of the skin, can be treated locally with ice baths and ointments. Second-degree burns, characterized by formation of blisters, may require the care of a doctor to prevent infection. Third-degree burns, with destruction of upper and lower Skin layers, are serious and often fatal; they require prompt medical attention to reduce pain and prevent Shock and infection. Blood Transfusion may be necessary to replace body fluids; long-term treatment may include Transplantation of natural or artificial skin grafts.

Bursitis. Acute or chronic inflammation of a bursa, a fluid-filled sac located close to a joint. Caused by infection, injury, or diseases like Arthritis and Gout, the inflammation produces pain, tenderness, and restricted motion. Treatment includes rest, application of heat, use of Analgesics and other drugs, and occasionally surgery.

Bush, Vannevar. 1890-1974, American electrical engineer and physicist; b. Everett, Mass. While professor and dean of engineering at the Massachusetts Institute of Technology (1923-38), Bush designed the differential analyzer, an early computer. During World War II, he led the U.S. Office of Scientific Research and Development, directing such programs as the development of the first atomic bomb.

Butadiene. A colourless flammable gas. $H_2C:CHHC:CH_2$, manufactured by the catalytic dehydrogenation of butanes and butanes. It polymerizes readily and is used in making synthetic rubbers. Butadiene is a conjugated diene and can undergo the Diels-Alder reaction. M.pt. -109°C; b.pt. -4.5°C; r.d. 0.62 (liquid).

Butanal Butyraldehyde). A colourless flamable liquid. CH_3 $(CH_2)_2$ CHO. M.pt -99°C: b.pt. 757°C; r.d. 08.

Butane. A colourless flammable gaseous alkane, C_4H_{10}, present in crude petroleum. It is used as a fuel, especially when liquefied under pressure in cylinders, and as raw material for the manufacture of synthetic rubber. Butane is sometimes called *n-butane* to distinguish it from isobutane. M.pt. -138.4°C; b.pt. -0.5°C; r.d. 0.60 (0°C).

Butanedioic acid. *See* succinic acid.

Butanoic acid. *See* butyric acid.

Butanol. Either of two isomeric straight-chain alcohols with the formula C_4H_9OH. *Butan-1-ol (n-*butyl alcohol) is a colourless flammable liquid, $CH_3(CH_2)_2CH_2OH$, manufactured from butanal and used as a solvent. M.pt. -89.5° C; b.pt. 117.3°C; r.d. 0.81. *Butan 2-ol (sec-*butyl alcohol) is also a colourless flammable liquid, $CH_3CH_2CH\text{-}OHCH_3$. *See also* butyl alcohol.

Butanone Mythyl ethyl ketone). A colourless flammable liquid ketone, $CH_3COC_2H_5$, manufactured by the oxidation of butane and used in lacquers and cleaning fluids. M.pt. -86.4°C; r.d.0.8.

Butenedioic acid. Either of two geometrical isomers with formula HCOO-CH:CHOOCH, maleic acid and fumaric acid.

Butte. An isolated hill with steep sides and a flat top, resulting from the more rapid erosion of the surrounding areas. Many occur in the plains of the W U.S. See Mesa.

Butter. Dairy product obtained by churning milk or cream until its fat solidifies. Cow's milk is generally the basis for butter, but milk from goats, sheep, and mares has also been used. Butter was known by 2000 B.C., and as it became a staple food various kinds of_hand churns were developed. Traditional butter-making involves cooling milk in pans, allowing the cream to rise and skimming it off, letting the cream ripen by natural fermentation then churning it. Farm-made until 1850, butter has since then increasingly become a factory product. Commercially made butter usually contains 80% to 85% milk fat, 12% to 16% water, and 2% salt. Butter without salt is known as sweet butter. The world's leading butter producers are the USSR, France, West Germany, the U.S., and India. Clarified butter (butterfat with milk solids removed) is widely used in Egypt and in India, where it is known as ghee. The high dietary value of butter is due to its large proportion of easily digested fat and to its vitamin A and vitamin D content.

Butter-and-eggs. Perennial plant *(Linaria vulgaris)* of the figwort family, originally native to Europe but now common in the U.S. It has small, snapdragonlike flower of yellow and orange and is consequently also known as wild snapdragon. Other yellow and orange flowers are sometimes called butter-and-eggs.

Buttercup or Crowfoot. Common name for the Ranuculaceae, a family of chiefly annual or perennial herbs of cool regions of the Northern Hemisphere. Primitive plants, they typically have a simple flower structure. The buttercups and crowfoots comprise the largest genus *(Ranunculus)*. Found mainly in the arctic, north temperate, and alpine regions, most have glossy yellow flowers and deeply cut leaves. A dozen or more species are common in every part of the U.S. The family also includes many other wildflowers and cultivated ornamentals. Some of these are the Anemone, Clematis, Columbine, Larkspur, and Peony.

Butterfly. Flying Insect that with the Moth comprises the order Lepidoptera. Butterflies have coiled, sucking mouthparts; two pairs of wings that function as one; and antennae with knobs at the tips. Most feed on nectar from flowers and are active by day. The butterfly larva (caterpillar) is transformed into a pupa (chrysalis) with a hardened outer integument within which it changes into the adults of most species live only about a month.

Butter of antimony. *See* antimony chloride.

Butter of zinc. *See* zinc chloride.

Butyl group. A monovalent group C_4H_9-, derived from butane by removing one atom of hydrogen. The term is often used for the group with a straight chain or carbon atoms: i.e. for the *n-butyl group*. There are three other isomers: the *isobutyl group*, $(CH_3)_2CHCH_2$-: the *sec-butyl group* $CH_3CH_2CH(CH_3)$—; and the *tert-butyl group*, $(CH_3)_3C$—.

Butyl rubber. A type of synthetic rubber made by copolymerization of isobutylene and small amounts of isoprene. Butyl rubber can be vulcanized with sulphur. It has a much lower permeability to gases than natural rubbers and is used in tyre inner tubes.

Butyraldehyde. *See* butanal.

Butyric acid (butanoic acid). A colourless liquid fatty acid, CH_3CH_2COOH, with a rancid odour, present, in the form of its esters, in bitter. Its main use is in making esters for perfumes and flavourings B.pt. 163°C; r.d. 0.96.

Buzzard. Hawk of the genera *Buteo* and *Pernis*. Honey buzzards (*pernis)* feed on insects, wasp and bumblebee larvae, and small reptiles. The term *buzzard* is often incorrectly applied to various hawks and new World Vultures.

C. Chemical symbol of the element Carbon.

Ca. Chemical symbol of the element Calcium.

Cabbage. Leafy garden vegetable of many widely dissimilar varieties, all probably descended from the wild, or sea, cabbage *(Brassica oleracea)* of the Mustard family, found on European coasts. Cabbage is used as food for humans and animals. Varieties include Broccoli, Brussels Sprouts, Cauliflower, Collards, Kale, and kohlrabi. All grow best in cool moist climates, chinese cabbage is a separate species.

Cable. Usually, high-strength wire cord or heavy metal chain used for hauling, towing, supporting the roadway of a suspension bridge, or securing a large ship to its anchor or mooring. Electric cables are conductors used for the transmission of electrical signals. A coaxial cable, virtually immune to external electomagnetic interference, consists of a tube of copper or other conducting material, the center of which contains another conductor. The two conductors are separated by an insulator. By means of coaxial cables a large number of telegraph and telephone messages, as well as television images, can be simultaneously transmitted. An intertwined and insulated group of wires that conducts electricity from generator to consumer is called a cable.

Cable television (CATV). Form of Broadcasting that sends programs to paying subscribers by means of coaxial Cable rather than through the airwaves. It originated (1949) to serve areas where reception was poor or nonexistent, and expanded rapidly in the mid-1970s with the easing of government restrictions and the use of Communications Satellite transmission. Further programming and audience growth came in the early 1980s, when the government moved to deregulate the industry. In addition to providing the programming of the local television broadcasting stations, CATV operators often offer additional channels of first-run movies, sports events, and around-the-clock news programming. The most advanced systems offer two-way, or interactive, communication; information services such as Videotex; and burglar—and fire-alarm protection.

Cacao. Tropical tree (*Theobroma cacao)* of the sterculia family, native to South America and now widely cultivated in the Old World. The fruit is a pod that contains a sweet pulp with rows of embedded seeds (the cocoa "beans" of commerce). Cocoa is obtained by fermenting the pods and then curing and roasting the extracted seeds. The resulting clean kernels, or cocoa nibs, are then processed. Chocolate is one product. Cocoa products have high food value due to their fat, carbohydrate, and protein content. Other uses are in cosmetics and medicines.

Cactus. Common name for the Cactaceae, a family of succulent plants found almost entirely in the New World. Cactus plants have fleshy green stems that function as leaves (the leaves are typically insignificant or absent), and, usually, spines of various colors, shapes, and arrangements. The large, showy, delicate flowers are commonly yellow, white, red, or purple. Cactus fruits are berries, some of which are edible. The reduced leaf surface and fleshy stem make cacti well fitted for water storage and retention. An extensive, ramified root system makes the plants adaptable to hot, dry regions, although cacti are not restricted to the desert. Most cacti blossom briefly in the spring, sometimes for only a few hours. Blossoms are especially sensitive to light, and different species bloom at different times of the day, e.g. the night-blooming cereus, whose fragrant blossoms unfold after sunset and last one night. Many species are cultivated for food or ornamentals; the hallucinatory drug Peyote comes from a cactus of that name.

Cadmium (Cd). Metallic elements, discovered in 1817 by Friedrich Stromeyer. Cadmium is a silver-white, lustrous, malleable, ductile metal. Its major use is as an electroplated coating on iron and steel to prevent corrosion.

Cadmium. Symbol: Cd. A soft bluish-white metallic element found associated with zinc ores, from which it is obtained as a by-product. It is used in bearing alloys and solder and in electroplated coatings for metal parts. Like zinc, it reacts with oxygen and other nonmetals and dissolves in nonoxidizing acids. Its compounds are similar to those of zinc, A.N. 48; A.W. 112.4; m.pt. 320.9°C; b.pt. 765° C; r.d. 8.65; valency 2.

Cadmium cell. *See* Weston Cell.

Cadmium sulphide. An insoluble yellow or orange solid. CdS, occurring naturally as *greenockite.* It is an important pigment (*cadmium yellow*) and is also used in transistors, fluorescent screens, and small batteries. M.pt. 1759°C (100 atm); r.d. 4.82.

Cadmium yellow. *See* cadmium sulphide.

Caesium. Symbol: Cs. A soft silvery-white metallic element found in some rare minerals. It can be made by electrolysis of the fused cyanide. Caesium is one of the most electropositive elements (second only to francium) and is used in photoelectric cells and similar devices. It is an alkali metal. A.N. 55; A.W. 132.905; m.pt. 28.5°C; b.pt. 690°C; r.d. 1.90; valency 1.

Caesium clock. An apparatus used for obtaining the steady frequency used in the definition of the second. The device depends on the fact that in a magnetic field the caesium atom can have two discrete energies due to the spin of its nucleus. A beam of atoms is passed into a region of strong nonuniform magnetic field in which the atoms are separated according to their energy state (*see* Stern-Gerlach experiment). Atoms with the lower energy are allowed to pass through a slit into a region in which there is a uniform magnetic field and a source of radiofrequency radiation with a frequency of 9 192 631 770 hertz. This excites some of the atoms to the higher energy state (*see* nuclear magnetic resonance). The beam is then passed into another nonuniform field in which atoms in the lower state are deflected onto a detector. The signal from this detector is fed to the radiofrequency supply and used to stabilize it on the particular frequency required for the transition. The device is stable to one part in 10^{11}.

Caffeine. Odorless, slightly bitter Alkaloid found in coffee, tea, Cola nuts, Mate, and cocoa (*see* Cacao). In moderation, caffeine is a mild stimulant that increases urination and the heart rate and rhythm. Excessive intake can cause restlessness, insomnia, heart irregularities, and delirium.

Calamine. 1. Mixture of zinc oxide with small amounts (0.5%) of iron III oxide, used to counteract inflammation of the skin.

2. Either of two zinc minerals: the carbonate, $ZnCO_3$. or a hydrated silicate, $2ZnO.SiO_2.H_2O$.

Calcination. The gradual deposition of calcium carbonate from hard water in pipes, kettles, etc.

Calcite. Widely distributed calcium carbonate mineral ($CaCO_3$) that ranges from white or colorless through a great variety of colors, owing to impurities. Its crystals are noted for their perfect cleavage. Calcite also occurs in a number of massive forms, including Marble, Limestone, and Chalk. Other forms include Iceland Spar, Stalactite and Stalagmite formations, Oriental Alabaster, and marl. It is used as a building stone and is the raw material for quicklime (Calcium Oxide) and Cement.

Calcite (Iceland spar). A colourless crystalline natural form of calcium carbonate, $CaCO_3$. Crystals of calcite show double refraction.

Calcium (Ca). Metallic element, first isolated in 1808 by Sir Humphry Davy. It is a silver-White, soft, malleable Alkaline-Earth Metal. The fifth most abundant element (about 3.6%) of the earth's crust, it is not found uncombined but occurs in numerous compounds, e.g., Apatite, Calcite, Dolomite, Iceland Spar, Limestone, and Marble. Calcium acts as a reducing agent in the preparation of other metals. It occurs in most plant and animal matter, and is essential for the formation and maintenance of strong bones and teeth. Calcium helps to regulate the heartbeat and is necessary for blood clotting. *See* Element (table); Periadic Table.

Calcium. Symbol: Ca. A soft white metallic element occurring extensively in nature, especially as forms of calcium carbonate (limestone, marble, chalk, etc.). Calcium is made by electrolysis of fused calcium chloride and is used as a reducing agent and getter. It is an alkaline-earth metal. A.N. 20; A.W. 40.08; m.pt. 850 C; b.pt. 1440°C; r.d. 1.58; valency 2.

Calcium carbide (carbide). A grey solid, CaC_2, made by heating limestone with coke. It reacts with water to produce acetylene. R.d. 2.22.

Calcium carbonate. A white powder or colourless crystalline solid, $CaCO_3$, occurring naturally as aragonite, calcite, chalk, limestone, and marble. In a precipitated form it is used as a gastric antacid and in toothpowders, flour, and white polishes. R.d. 1.71.

Calcium chloride. A white crystalline or granular deliquescent solid, $CaCl_2$, used as a preservative and a drying agent. It can also be obtained as the monohydrate, dihydrate, and hexahydrate. M.pt. 772° C; r.d. 2.2 (anhydrous).

Calcium cyanamide (cyanamide). A grey-black powder, $CaCN_2$, manufactured by heating calcium carbide in nitrogen. The reaction is a method of nitrogen fixation: cyanamide is hydrolysed to ammonia and acetylene and is used as a fertilizer. M.pt. 1300° C; r.d. 1.08.

Calcium cyanide. A highly prisonous colourless or grey solid, $Ca(CN)_2$. used as a plant insecticide and rat poison. Decomposes at 350°C.

Calcium cyclamate. A white powder, $(C_6H_{11}NHSO_3)_2Ca.2H_2O$, with a sweet taste, formerly used as a sweetener in soft drinks.

Calcium fluoride. A white powder, CaF_2, occurring naturally as fluorspar. It is the main source of fluorine and its compounds. M.pt. 1360°C; b.pt. 2500°C; r.d. 3.2.

Calcium hydrate. *See* calcium hydroxide.

Calcium hydroxide (lime, slaked lime, calcium hydrate). A soft white powder, $Ca(OH)_2$, prepared by the action of water on calcium oxide. It is used in cement and in neutralizing acid soil. R.d. 2.24.

Calcium nitrate. A deliquescent crystalline solid, $Ca(NO_3)_2$, used in fertilizers and explosives. M.pt. 561°C; r.d. 2.5.

Calcium oxide or **calcia.** Chemical compound (CaO), also called lime, quicklime, or caustic lime. A colorless crystalline or white amorphous substance, it has wide industrial uses, e.g., in making porcelain and glass; in purifying sugar; in preparing bleaching powder, calcium carbide, and calcium cyanamide; in water softeners; in mortars and cements; and in treating acidic soil (liming).

Calcium oxide (lime, quick lime, calx). A white or grey granular solid, CaO, made by roasting natural forms of calcium carbonate. It is widely used, chiefly in making calcium hydroxide, as a flux in steel manufacture, in making paper, and in water softening. M.pt. 2580°; b.pt. 2850°C; r.d. 3.40.

Calcium phosphate. Any calcium salt of a phosphoric acid. Calcium phosphates are found in some rocks and are also present in bone. They are used as fertilizers (*see* superphosphate). The term is often used for calcium orthophosphate, $Ca_3(PO_4)_2$, which is a white insoluble powder. M.pt. 1670°C; r.d. 3.14.

Calcium sulphate. A white powder, $CaSO_4$ or $CaSO_4.2H_2O$, occurring naturally as anhydrite or gypsum. Calcium sulphate hemihydrate, $CaSO_40.5\text{-}H_2O$, is plaster of pairs. M.pt. 1450°C; r.d. 2.96.

Calcium sulphide. A colourless crystalline solid, CaS, made by reducing calcium sulphate with carbon at high temperature. It is a luminescent compound. R.d. 2.5.

Calculator, electronic. Electronic device for performing numerical computations. Electronic calculators became available in the early 1960_s, and in the early 1970_s miniature types, some of them pocket size, were marketed as consumer items. Electronic calculators have ten keys that can be used to enter numbers into the machine; additional keys are provided to enable the user to perform a range of operations, from basic arithmetic in simple devices to the generation of complex mathematical functions in more advanced types. The result of an operation are either shown on an electronic display or printed. Some of these machines are actually small Computers with limited memory programming capabilities.

Calculus. The branch of mathematics concerned with integration and differentiation of functions. It is divided into the integral calculus and the differential calculus.

Calculus. Branch of Mathematics that studies continuously changing quantities. It was developed in the 17th cent. independently by Sir Isaac Newton and G.W. Leibniz. The calculus is characterized by the use of infinite processes, involving passage to a Limit. The *differential calculus* arises from the study of the rate at which a Function, usually symbolized by *Y or f(x),* changes relatives to a change in the independent variable, usually *x*. This relative rate can be computed from a new function—the derivative of *Y* with respect to *x*, denoted by dy/dx, Y', or $f'(x)$—arrived at by a process called *differentiation*. Formulas have been developed for the derivatives of all commonly encountered functions. For example,if $Y=X^n$ for any real number *n* except -1, then $Y' = nx^{n-1}$, and if $Y = \sin x$, then $Y' = \cos x$ (*see* Trigonometry). In physical applications, the independent variable is frequently time, e.g., if $s = f(t)$ expresses the relation between the distance *s* traveled and the time *t* elapsed, then $s' = f'(t)$ represents the rate of change of distance with time, i.e., the speed or velocity (*see* Motion) at time *t* Geometrically, the derivative is interpreted as the slope of the line tangent to curve at a point. This view of derivative yields application, e.g., in the design of optical mirrors and lenses and the determination of projectile paths. The *integral calculus* arises from the study of the limit of a sum of elements when the number of such element increases without bound while the size of the element diminishes. Conventionally, the area *A* under the curve $y = f(x)$ between the two values $x = a$ and $x = b$ is symbolized by $A = \int_a^b f(x)\,dx$, called the definite integral of $f(x)$ from *a* to *b*. The area is approximated by summing the products of $f(x)$ and dx for each of the infinitely small distances (dx) that comprise the measurable distance between *a* and *b*. This method can be used to determine the lengths of curves, the areas bounded by curves, and the volumes of solids bounded by curved surfaces. The connection between the integral and the derivative is known as the Fundamental Theorem of the Calculus, which, in symbols, is $\int_a^b f(x)dx = F(b)\ F(a)$, where $F(x)$ is a function whose derivative is $f(x)$. The calculus has been developed to treat functions not only of a single variable but also of several variables and is the foundation for the larger branch of mathematics known as Analysis.

Calendar. System if reckoning time usually based in a recurrent natural cycle, such as the cycle of the sun through the seasons or the moon through its phases. Because the solar year is 365 days 5 hr 48 min 46 sec and the lunar year (12 synodic months of 29.53 days) is 354 days 8 hr 48 min, people have been confronted from ancient times with the problem of the discrepancy. Because the year is not exactly divisible by months and days, the practice arose of making arbitrary divisions and inserting extra (inctercalary) days or months. The Gregorian calendar, generally accepted today, evolved from the Roman calendar reformed (46 B.C.) by Julius Caesar. In the Julian calendar April, June, September, and November had 30 days, February 28 days (29 days every fourth, or leap, year), and all other months 31 days. The date was computed by counting backward from the Kalends (the 1st day), the Nones (the 7day in March, May, July, and October; the 5th day in order months), and ides (the 15th day in March, May, July, and October; the 13th day in other months); hence Jan. 10 was the 4th day of the ides of January. Because the julian year of 365 days 6 hr was too long, by the 16th cent. the vernal equinox was displaced from March 21 to March 11. Pope Gregory XIII ordained that 10 days be dropped in 1582 and that years ending in hundreds be leap years only if divisible by 400. The non-Roman Catholic countries were slow to accept the Gregorian (New Style) calendar; it was adopted in England in 1752 and by the Eastern Church in the 20 cent. The Christian ecclesiastical calendar was based on the belief that Jesus' resurrection was on a Sunday, hence easter should fall on Sunday. The First Council of Nicaea (325) decreed that Easter be the Sunday, following the first full moon after the vernal Equinox; today the date varies of the lunar period are not considered. Other calendars include the Jewish calendar (12 months, plus intercalary months 7 times a 19 years) and the Muslim lunar calendar.

Calibration. The process of using an instrument or apparatus to take readings of standard or known quantities so that the arbitrary readings given by the instrument can be converted into absolute values.

Caliche. *See* sodium nitrate.

Californium. Symbol: Cf. A transuranic actinide element made, in trace amounts, by neutron bombardment of berkelium. The most stable isotope, ^{294}Cf, has a half-life of 470 years. A.N. 98.

Calligraphy [Gr., = beautiful writing], skilled penmanship practiced as a fine art. Papyrus fragments from the 4th cent. B.C. show two early European types of handwriting: a cursive script, used for letters and records, and a more polished style, called uncials, used for literary works. This style was developed after the first cent. A.D., giving rise to many splendid scripts throughout the Renaissance. With the establishment of the printing press, the use of calligraphy declined until the 20th cent., when interest in the art was revived by the work of Owen Jones and William Morris. Calligraphy has always been valued by Far Eastern peoples as a major aesthetic expression, equal to or more significant than painting. Fine examples of islamic calligraphy are found in many beautiful copied and decorated Korans.

Calomel electrode. A half cell consisting of a mercury electrode covered with a paste of mercury I chloride and mercury in a solution of potassium chloride saturated with mercury I chloride. The system is used as a standard electrode: when the potassium chloride solution is satruated its potential is -0.2415 volts at 25°C with respect to a hydrogen electrode.

Caloric theory. A theory concerning the nature of heat, in which heat is considered to be a weightless fluid *(caloric).* The theory was abandoned in the mid-nineteenth century when experiments on the mechanical equivalent of heat showed that heat is a form of energy.

Calorie (Cal), unit of energy required to raise the temperature of one gram of water one degree Centigrade (from 14.5° to 15.5°C); 1 cal = 4.1840 Joules. Nutritionists use the kilocalorie (1,000 cal) to state the heat content of food.

Calorie (international table calorie). A unit of heat energy equal to 4.1868 joules. Formerly, the calorie was defined as the amount of heat required to raise the temperature of one gram of water through one degree centigrade. This value depends on the temperature of the water, leading to considerable confusion in the past about the use of the calorie. The *fifteeh-degree calorie or small calorie* is the heat required to raise one gram of water from 14.5°C to 15.5°C at standard pressure. It is equivalent to 0.999 690 international table calorie.

Calorific value. A measure of the quality of a fuel, equal to the amount of heat produced by complete combustion of unit mass of the fuel.

Calorimeter. Any apparatus for making thermal measurements in calorimetry.

Calorimerty. The measurement of heat as used in the determination of heat capacities, latent heats, calorific values, heats of combustion, and similar quantities.

Calx. 1. A metal oxide formed by roasting an ore.

2. *See* calcium oxide.

Cambium. Thin layer of reproductive tissue lying between the Bark and the wood of a Stem, most active in woody plants. In herbaceous plants the cambium is usually inactive; in monocotyledons it is absent. Producing thin layers of phloem on the outside and xylum on the inside, cambium growth increases the diameter of the stem. Its seasonal growth is responsible for Annual Rings.

Camera. 1. An apparatus for producing a photographic image. It consists of a light-tight box with a lens to focus a real image onto photographic film.

2. (television camera). A device for converting an optical image into electrical signals. Essentially it consists of a lens system that focuses the image onto a screen consisting of a mosaic of light-sensitive particles. An electrical signal is obtained, proportional to the light falling on each element of the mosaic, by using photoconduction or photoemision.

Camel. Hoofed ruminant (family Camelidae). The family consists of the true camels of Asia, the wild guanaco and domesticated Alpaca and LLAMA of South America, and the vicuna of South America. The two species of true camel are the singal-humped Arabian came, or dromedary *(Camelus dromedarius),* a domesticated animal of Arabia and N. Africa; and the two-humped Bactrian camel *(C. bactrianus)* of central Asia. Their humps are storage places for fat. Ranging in color from dirty white to dark brown, camels are well adapted for desert life and can go without water for several days.

Camellia. Evergreen shrub or small tree (genus *Camellia)* of TEA family, nativse to Asia, now widely cultivated for the white, red, or variegated showy blossoms and glossy, dark-green foliage. Tea-seed oil, from the seeds of *sasanqua,* is used in cooking, and in soap and textile manufacturing.

Cameo. Small relief carving, usually on striated precious or semiprecious stones or on shell. The design, often a portrait head, is cut in the light-colored vein; the dark vein becomes the background. Glass of two colors in layers may also be cameo-cut. The art originated in Asia and spread to ancient Greece and Rome; it was revived during the Renaissance and in the Victorian era.

Camera. In Photography, device for recording an image on film or some other light-sensitive material. It consists of a lightproof box; a Lens through which light enters and is focused; a shutter that controls the size of the lens opening and the length of time it is open; a mechanism for moving the film between exposures; and a viewfinder, or eyepiece, that shows the user the image the lens sees. The camera developed from the Greek *camera obscura* [Lat., = dark chamber], an artist's tool dating from the Middle Ages. It was a light-tight box with a convex lens at one end and a screen that reflected the image at the other; the artists traced the image. Joseph Niphore Niepce produced the first negative image n 1826.

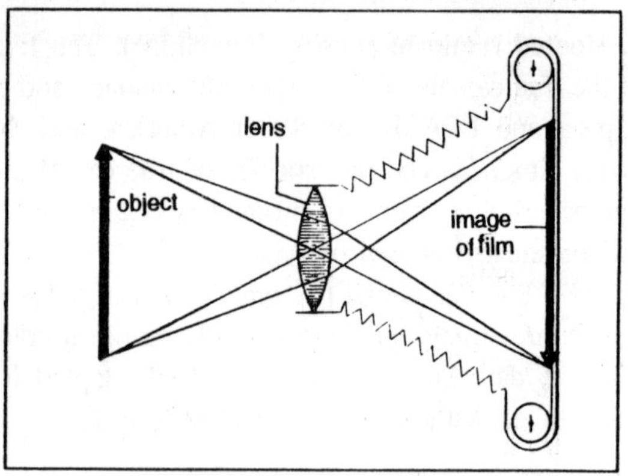

Fig. C-1. Image formed by a camera.

Camphor ($C_{10}H_{16}$). White, crystalline solid with a pungent odor and taste. It can be obtained from the camphor tree or synthesized from oil of Turpentine. Camphor is used to make Celluloid and lacquers. In medicine it is used as a stimulant, a diaphoretic, and an inhalant; camphor ice, a mixture chiefly of camphor and Wax, is applied externally. The alcoholic solution is known as spirits of camphor.

Camphor. A colourless crystalline ketone, $C_{10}H_{16}O$. obtained from the wood of the camphor tree. It is used as a plasticizer for celluloid and as an insect repellent. M.pt. 179°C; b.pt. 204°C; r.d. 0.99.

Canada balsam. A transparent resin obtained from certain coniferous trees, used as a cement in optical systems because its refractive index is similar to that of glass.

Canal. An artificial waterway constructed for navigation or for the movement of water, e.g., irrigation, drainage of wetlands, and flood control. Irrigation canals probably date back to the beginnings of agriculture. Navigation canals developed later and for a long time were level, shallow cuts or had inclined planes up which vessels were hauled from one level to the next. The *carnal lock,* a stretch of water enclosed by gates at each end, provides another means of raising or lowering vessels. When a vessel enters a lock through one gate, the gate is closed behind it. Water is let into or drained out of the lock until its level equals that of the water ahead. The vessel then passes out through the forward gate. Canal locks developed separately in China (10th cent). and Holland (13th cent). The Grand Canal of China is the world's longest canal. Europe developed a complex system of canals, and in the U.S. canal-building boomed after the opening (1825) of the Erie Canal. The rise of rail-roads, however, brought a decline in the use of canals as inland waterways. Canals have also been built to shorten sea routes, e.g., the Panama Canal and the Suez Canal.

Canal rays. Streams of positive ions obtained from a discharge tube by boring small holes in the cathode. Some of the ions moving to the cathode pass through forming positive rays. They were first obtained by Sir J.J. Thomson in his experiments on charge-to-mass ratios.

Canary. Small Bird of the Finch family, descended from either the wild serin finch or the wild canary (*Serinus canarius*) of the Canary Islands, Madeira Island, and Azores. The wild birds usually grey or green; breeding has yielded plain and variegated birds, mostly yellow and buff. Captive canaries are trained to sing and can live for 15 years or more.

Cancer. Common term for tumors (neoplasms) characterized by uncontrolled growth. Unlike normal cells, cancer cells are atypical in structure do not have specialized functions. They compete with normal cells for nutrition, eventually killing normal tissue by depriving it of nutrition, Cancerous, or malignant, tissue can remain localized, invading only neighboring tissue, or can spread to other tissues or organs via the Lymphatic System or blood (i.e., metastasize); virtually all tissues and organs are susecptible. Cancer symptoms, which are often nonspecific, include weakness and loss of appetite and weight. The causes are thought to include chemical agents that alter the cell's nucleic acids, e.g., nitrogen mustard; toxic substances, e.g., Asbestos, pollutants, substances in cigarette smoke; hormonal imbalances, possibly brought on by drugs, e.g., Des; genetic tendencies for certain types of cancer, e.g., of the breast or stomach; physical agents such as X rays and radioactivity; and Viruses. Methods of detection include visual observation, palpation, X-ray study, Cat Scans, Ultrasound, and Biopsy. Tumors caught early, before metastasis, have the best cure rates. Cancer is treated by surgery, Chemotherapy, and Radiation. The branch of medicine concerned with diagnosis, treatment and research into the causes of cancer is known as oncology.

Candela. Symbol: cd. The base SI unit of luminous intensity, equal to the luminous intensity of 1/600 000 sq. metre of the surface of a black body at a temperature of 2040 kelvins and a pressure of 101 325 pascals. The temperature is the freezing point of platinum.

Candle. Cylinder of wax or tallow containing a wick, used for illumination, for ceremonies (primarily religious), or as ornamentation. Used in ancient times, candles became common in Europe by the Middle Ages. Tallow, beeswax, and vegetable wax were supplemented by spermaceti in the late 18th cent., stearine c. 1825, and paraffin c.1850. Originally handmade by dipping, molding, or pouring, candles are now generally machine-made. They are often scented.

Candle A former unit of luminous intensity. The *international candle* was defined in terms of the luminous intensity of the Harcourt pentane lamp. The unit was replaced by the candela in 1948.

Candlepower. Luminous intensity as measured in candles.

Candytuft. Low-growing old World plant (genus *Iberis)* of the Mustard family, often cultivated as a rock garden and border ornamental. Candytufts have flat-topped or elongated clusters of variously colored flowers.

Cane. In botany, name for a hollow or woody, usually slender, jointed plant stem (e.g., Rattan and some bamboos) and for some tall Grasses (e.g., Sugarcane and Sorghum). Giant cane *(Arundinaria mascrosperma or gigantea),* Bamboo grass native to the U.S., often forms impenetrable thickets, the canebrakes of the South.

Cannibalism. Practice of eating human flesh, found in Africa, South America, the South Pacific islands, and the West Indies. Victims for rites of life-or power-transfer are usually sought among alien groups, although some peoples eat part of a kinsman's corpse as a gesture of respect for the deceased. Headhunting may have evolved from cannibalism.

Canning. Process of hermetically sealing cooked food for future use, including meat, poultry, fruit and vegetables, seafood, milk preserves and pickles, jams and jellies. Developed early in 19th-cent. France by Nicolas Appert, a chef, the method spread to other European countries and to the U.S., where it was patented around 1815. The process was widely used to provide foodstuffs to soldiers during the American Civil War. The early glass and tin containers were ultimately supplanted by tin-plated steel cans, mass-produced in America since 1847 and now the basis of a huge modern industry. In home as in factory canning, the process involves cleaning and preparing the raw product (with rapid handling necessary to prevent vitamin loss, bacterial spoil-age, or enzyme changes). Once filled, the containers are thermally exhausted to release gases, then subjected to heat to destroy any microorganisms.

Cannizzaro, Stanislao. (Kannettsa'ro), 1826-1910, Italian chemist. professor at Palermo and then at Rome, he is known for his discovery of cyanamide and for his method of synthesizing alcohols (Cannizzaro's reaction). He also explained how atomic weights could be determined on the basis of Avogadro's law and distinguished from molecular weights.

Canonical. Denoting or involving a general mathematical rule or formula.

Canonical equations. A set of equations used in classical mechanics to describe the motion of a set of particles. They have the form $dp_l/dt = \delta H/\delta q_i$ and $dq_l/dt = H/\delta p_i$. H is the Hamiltonian function, *q* the position coordinate, and *p* the momentum.

Capacitance. Symbol: *C* The ability of a system to store electric charge or a measure of this for a particular system, equal to the charge *(Q)* stored to rise the system's potential *(V)* by one unit (i.e. $C = Q/V$). The unit of capacitance is the farad.

Capacitance. In electricity, the capability of a body, a system or an Electric circuit for stroing electric Charge. Capacitance, in units of farads, is expressed as the ratio of stored charge in coulombs to the applied potential difference in volts. In electric circuits, devices designed to store charge are called Capacitors. When alternative current flows through a capacitor, the capacitor produces a reactance, inversely proportional to the capacitance, that resists the current flow.

Capacitor or **condenser.** Device for storing electric charge. simple capacitors usually consist of two plates made of an electrically conducting material (e.g., glass, paraffin, mica, oil, or air). If an electric Potential (voltage) is applied to the capacitor plates, the plates will become charged, one positively and one negatively. If the externally applied voltage is then removed, the capacitor plates remain charged, and the electric charge induces an electric potential between the two plates. This phenomenon is calledelectrostatic Induction. This capacity of the device for storing electric charge (i.e., its capacitance) can be increasing the area of the plates, by decreasing their separation, or by varying the substance used as an insulator. The Dielectric constant is a measure of the increase in capacitance due to a particular insulator used to separate the plates. The Leyden jar, a form of capacitor invented at the Univ. of Leiden in the 18th cent., consists of a narrow-necked glass jar coated on part of its inner and outer surfaces with conductive metal foil.

Capactior (condenser) A device consisting of conductors and insulators with the property of storing electric charge. In the *parallel-plate capacitor* two parallel metal plates are separated by air or some other insulator (the dielectric). If a potential difference V is applied across the plates, each plate acquires a charge Q. The capacitance is Q/V, and is equal to ${}_{E}A/d$, where A is the area of the plates, d the distance between them, and E the permittivity of the material between the plates.

Capacitors are used to introduce reactance into alternating-current circuits. The usual fixed type is constructed of layers of waxed paper. An *electrolytic capacitor* has two aluminium electrodes separated by paper soaked in electrolyte or by liquid electrolyte. The dielectric is an insulating oxide layer on the anode. In this type one electrode must always have a positive potential with respect to the other. Variable capacitors have a set of moving plates and a set of fixed plates, so that the effective area can be changed.

Caper. Common name for members of the family Capparidaceae, chiefly Old World tropical plants closely related to the Mustard family. The pickled caper, used as a condiment, is the flower bud of *Capparis spinosa,* cultivated in the Mediterranean area. The family also includes the spider flower *(Cleome spinosa)*, a common garden annual.

Capillarity or capillary action. Phenomenon in which the surface of a liquid is elevated or depressed when it comes in contact with a solid. The result depends on the outcome of two opposing forces, Adhesion and Cohesion. Adhesion between glass and water causes the water to rise along a glass wall until this force is balanced by the cohesive force acting to minimize the liquid's surface area *(see* Surface Tension). When adhesion is less than cohesion, as with glass and mercury, the surface is lowered. The upward flow of water in soil and in plants is partially caused by capillarity.

Capillary. Any long narrow hole or pore. A *capillary tube* is a tube with a narrow bore, especially a glass tube with thick walls, of the type used in mercury thermometers.

Capillary action (capillarity). The movement of liquid through capillary tubes, small pores, etc., as caused by surface tension.

Caproic acid. *See* hexanoic acid.

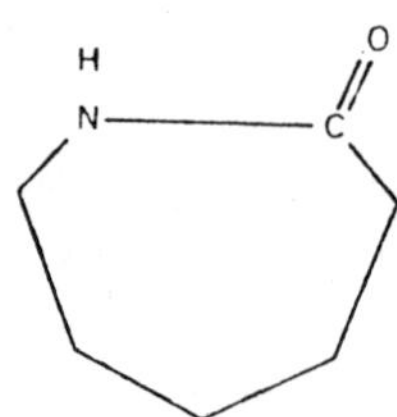

Fig. C-2. Caprolactam.

Caprolactam. A white crystalline powder, $C_6H_{11}NO$, used in the manufacture of nylon. M.pt. 70°C; b.pt. 15°C; r.d. 1.02.

Capture. Any process in which an additional particle is acquired by an atom, on, molecule, or atomic nucleus. The term is more usually applied to nuclear processes, as in the capture of a neutron by a nucleus, often followed by emission of a gamma ray *(radiative capture)*. Another nuclear process is *K-capture* in which the nuclear process is an orbiting electron from its K-shell. In the nucleus, a proton is changed into a neutron with associated emission of a neutrino. The atom is left with a vacant state in its K-shell which is filled by an electron from an outer shell with associated emission of an X-ray photon.

Carat. 1. A unit of mass used for precious stones, equal to 2 X 10 ^{10} kilogram. Formerly the carat was an Imperial unit of troy measure equal to 4 grains.

2. A measure of the quality of gold, equal to the number of parts of gold (by weight) in 24 parts of gold alloy. Thus 24 carat gold is pure gold, 22 carat gold is 22 parts gold and 2 parts other metal, etc.

Carbanion. A negatively charged organic ion in which the negative charge occurs on a carbon atom. An example is the ion $(C_6H_5)_3C$, which is found in the ionic compound triphenylmethyl sodium. The cycolpentadienyl ion, $C_5H_5^-$, is another example of a stable carbanion: in both these examples the ion is stabilized by resonance. Many carbanions exit only as transient intermediates in chemical reactions.

Carbene. A divalent species with the general formula RR'C:, where R and R' are atoms or groups. The simplest example is methylene, H_2C:, which can be produced by the photolysis of ketene or by heating diazomethane. Carbenes can exist in two states; one in which the two electrons are paired in the same atomic orbital and the other in which the electrons occupy different orbitals. The latter species is a *diradical.* All carbenes are extremely reactive, having a transient existence as intermediates in certain chemical reactions. They add to double bonds to form a three-membered cyclopropane ring. They are also able to insert themselves into C-H bonds: methylene for example produces $C\text{-}CH_3$.

Carbide. Any compound of carbon with a more electropositive group or element. Carbides are often made by combination of the elements at high temperatures (2000°C). Three main types can be distinguished.

Interstitial carbides are formed by iron, cobalt, nickel, and number of other transition metals. The carbon atoms occupy holes in the metal lattice and the substance has properties similar to those of the metal.

Covalent carbides are formed by silicon and boron. Both compounds are hard high-melting solids in which covalent bonds hold the atom in the lattice.

Ionic carbides are salts. Two types are known. The *methanides* appear to contain C^{4-} ions and yield methane on hydrolysis. An example is aluminium carbide Al_4C_3. The *acetylides* contain the ion C : C;, derived from acetylene. They give acetylene on hydrolysis. Calcium carbide, C_aC_2, is a common example and the term *carbide* is often used to refer to this compound.

Carbinol. A former name for methanol. The term was also used to name other alcohols as substituted derivatives of methanol. Ethanol, for example, was *methylcarbinol.*

Carbocyclic. Denoting a chemical compound whose molecules contain a ring of carbon atoms.

Carbodiimide *See* cyanamide.

Carbohydrate. Any of a class of compounds with the formula $C_nH_{2m}O_m$. The carbohydrates include the sugars and high-molecular-weight compounds such as starch and cellulose.

Carbohydrate. Any member of a large class of chemical compounds that includes sugars, starches, cellulose, and related compounds. Carbohydrates arc produced naturally by green plants from carbon dioxide and water. Important as foods, they supply energy and are used to make fats. The three main classes of carbohydrates are monosaccharides, which are the simple Sugars, e.g., Fructose and Glucose; disaccharides, which are made up of two monosaccharide units and include Lactose, Maltose, and Sucrose and polysaccharides, which are polymers with many monosaccharide units and include Cellulose, Glycogen, and Starch.

Carbolic Acid. *See* Phenol.

Carbon. (C), Nonmetallic element, known since ancient times. Pure carbon forms are amorphous carbon (found in such sources as Charcoal, Coal, Coke, Lignite, and Peat) and the crystals Graphite, a very soft, dark-grey or black, lustrous material, and Diamond, the hardest substance known. Organic Chemistry is the study of carbon compounds. All living organism contain carbon. Carbon has seven isotopes; carbon-12 is the basis for Atomic Weights; carbon-14, with a half-life of 5,730 years, is used to trace chemical reactions and to date geologic and archeologic specimens.

Carbon Symbol: C. A nonmetallic element present in all living matter, atmospheric carbon dioxide, and many mineral carbonates. There are two allotropic crystalline forms, graphite and diamond, and several amorphous forms, including gas carbon, charcoal, and lamp black. The element is widely used in electrodes, colloidal conductive coatings, crucibles, rubber inks, and carbon fibres. It reacts directly with other nonmetals at high temperatures. Metal carbides can also be made. Carbon is unique in its ability to form molecules containing long chains of atoms and over a million organic compounds are known. A.N. 6; A.W.12.011; m.pt. above 3500°C; r.d. 3.51 (diamond), 2.25 (graphite); valency 4.

Carbon cycle. Biology, the exchange of carbon between living organisms and the nonliving environment. Carbon, the central element in the compounds of which organisms are composed, is very derived from free carbon dioxide found in air or dissolved water. Plants incorporate carbon into carbohydrates and other complex organic molecules buy means of Photosynthesis; during Respiration, or Oxidation, they combine oxygen with portions of the carbohydrate molecule, releasing carbon in the form of carbon dioxide and water. Carbon is also returned to the environment when plants die and their organic materials broken down by bacteria and other microorganism. Animals obtain carbon by feeding on plants other animals; they release carbon through respiration and, after death, indirectly through the respiration of microorganisms that consume them.

Carbon-14. The radiocative isotope of carbon with mass number 14, present in small amounts in nature (*see* carbon dating). It is made by irradiation of a nitrate in a reactor and extensively used in tracer studies. Half-life 5720 years. *Carbon-14 dating.*

Carbonate. Any salt of carbonic acid.

Carbonation. The introduction of carbon dioxide into a liquid under pressure. Carbon dioxide is only slightly soluble in water forming a weakly acidic solution of carbonic acid. More CO_2 can be dissolved in this solution under pressure, and when the pressure is released the gas comes out of solution in tiny bubbles.

Carbon black. A fine powdered form of carbon made by incomplete combustion of hydrocarbons. It is used as a pigment in links and as a filler for rubber.

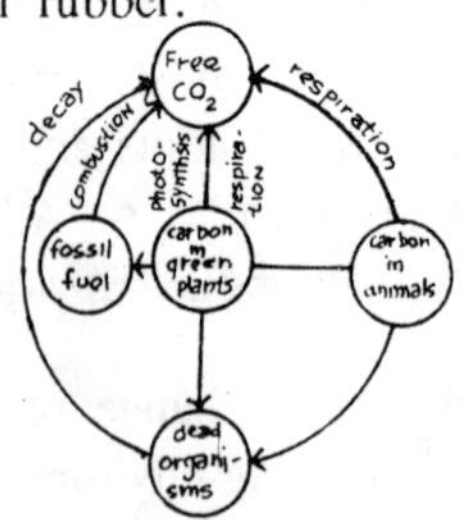

Fig. C-3. Carbon cycle.

Carbon cycle. A cycle of thremonuclear reactions production the energy in some stars. It results in the formation of helium nuclei from hydrogen nuclei.

Carbon dioxide (CO_2), Chemical compound, occurring a as colorless, odorless, tasteless gas that is about $1^1/_2$ times as dense as air under ordinary conditions. It does not burn and will not support combustion of ordinary materials. Its weakly acidic liquefied by compression and cooling, provides the sparkle in carbonated beverages. solid carbon dioxide, or dry ice, is a refrigerant. Dough rises because of carbon dioxide formed by the action of yeast and baking powder. Carbon dioxide is a raw material for Photosynthesis in green plants, and is a product of animal Respiration and of the decay of organic matter. Carbon dioxide occurs both free and combined in nature, and makes up about 1% of the volume of dry air. It can cause death by suffocation if inhaled in large amounts.

Carbon dioxide. A colourless odourless noncombustible gas, CO_2 produced by combustion of carbon compounds. It is also a product of respiration and is present in the atmosphere. Carbon dioxide is used in carbonated beverages, fire extinguishers, and for providing an inert atmosphere in welding. Solid carbon dioxide *(dry ice)* is used as a refrigerant. M.pt. -56.6°C (5.2 atm); sublimes at -78.5°C; r.d. 1.10 (37°). see *also* carbonic acid.

Carbon disulphide. A colourless poisonous flammable liquid, CS_2, which is almost odourless when pure but usually has a strong disagreeable odour caused by sulphur impurities. It is made from natural gas and sulphur. Carbon disulphide is used as a solvent and as a raw material for the manufacture of rayon and carbon tetrachloride. M.pt. -110.8°C; b.pt. 46.3° C; r.d. 1.26.

Carbonic fibres. Fibres of carbon made by heating fibres of rayon or other textiles. The material has an orientated crystal structure giving it a high tensile strength. It is used to reinforce strong light temprature-resistant material.

Carbonic acid. The weak disbasic acid, H_2CO_3, produced by dissolving carbon dioxide in water. It is only stable in aqueous solution and cannot be isolated. Its salts are the carbonates.

Carbonic acid (H_2CO_3), A weak dibasic acid formed when Carbon Dioxide dissolves in water; it exists only in solution. With bases it forms the Carbonate and bicarbonate salts.

Carbonium ion. A positively charged organic ion in which the positive charge occurs on a carbon atom. An example is the positive triphenylmethyl cation, $(C_6H_5)_3C^+$, which is stablized by resonance. Carbonium ions exist as short-lived intermediates in many chemical reactions.

Carbonize. 1. To treat with carbon.

2. (carburize). To change from an organic compound into elemental carbon, or to coat something with carbon by such a process. Normally, organic compounds are carbonized by incomplete oxidation in a restricted supply of air. Complete oxidation would lead to carbon dioxide and water.

Carbon monoxide. A colourless odourless flammable highly toxic gas, CO, produced by incomplete combustion of carbon compounds and manufactured by partial oxidation of coke or natural gas. Carbon monoxide is a neutral oxide. It burns with a blue flame and is used as a fuel and as a reducing agent in metallurgy. M.pt. -105°C; b.pt. -190° C; r.d. 0.97.

Carbon monoxide. Chemical compound (CO), colorless, odorless, tasteless, extremely poisonous gas that is less dense than air under ordinary conditions, it burns in air with a characteristic blue flame, producing carbon dioxide. It is a component of the artificial fuels producer gas and Water GAS. as a reducing agent, it removes oxygen from many compounds and is used in the reduction of metals from ores. When air containing as little as 0.1% carbon monoxide by volume is inhaled, the oxygen of hemoglobin is replaced by the carbon monoxide, resulting in fatal oxygen starvation throughout the body.

Carbon tetrachloride (tetrachloromethane). A colourless nonflammable heavy liquid, CCl_4, with a characteristic sweet odour. It is an unreactive compound, made by the chlorination of carbon disulphide or methane and used as a fire extinguishant, refrigerant, and solvent. M.pt.-23°C; b.pt.76.5° C;r.d. 1.59.

Carbonyl. Any of a class of transition metal compounds containing a metal atom coordinated to one or more carbon monoxide molecules. A common example is nickel carbonyl, $Ni(CO)_4$, which was used in the Mond process.

Carbonylation. Any chemical reaction that results in the production of a carbonyl group in a compound. An example is the dehydrogenation of propan-2-ol to acetone using a copper catalyst at 250°C: CH_3 $CHOHCH_3$ = CH_3COCH_3 + H_2.

Carbonyl group. The divalent group =CO, as present in aldehydes, ketones, carboxylic acids, and metal carbonyls.

Carborundum. ® silicon carbide used as an abrasive.

Carboxylation. Any chemical reaction that results in the production of a carboxyl group in a compound. The term is usually applied to the direct oxidation of an alkyl group to a carboxyl group, as in the reaction $ArCH_2CH_3$ = ArCOOH, where Ar is an aryl group.

Carboxyl group. The organic group -COOH, containing a carbonyl group linked to a hydroxyl group. It is the functional group in carboxylic acids.

Carboxylic acid. Any of a class of organic acids with the general formula R.CO.OH, where R is usually a hydrocarbon group. The strength of the acid depends on the nature of R: when it is a simple hydrocarbon group, as in acetic acid, the compound is usually a week acid. Carboxylic acids can be made by oxidation of the corresponding alcohol or aldehyde. They can be dehydrated to yield anhydrides and react with halogenating agents to form acid halides. They also show the usual reactions of acids, forming salts with alkalis and esters with alcohols. Carboxylic acids are often called *fatty acids,* owing to the fact that esters of many higher carboxylic acids are found in fats and oils.

Carburetor. Device in a gasoline engine that vaporizes the gas and mixes it with a regulated amount of air for efficient combustion in the engine cylinders. Land vehicles, boats, and light air craft have a float carburetor, in which a float regulates the fuel level in a reservoir from which the fuel is continuously sucked into the intake manifold at a restriction called a venturi. When there is an individual spray for metered spurt, the device is called a fuel injector.

Carbylamine reaction. A test for primary amines in which the sample is heated with an alcoholic solution of postassium hydroxide and chloroform. The amine is converted into a carbylamine (isocyanide) which is easily recognized by its unpleasant smell.

Cardinal or **redbird.** North American songbird of the FINCH family. The eastern cardinal *(Richmondena cardinalis)* male is bright red with a black throat and face; the female is brown with red patches. Both have crests and red bills.

Carius method. A technique for the quantitative analysis of sulpher and halogens in organic compounds. The sample is heated in a sealed tube with concentrated nitric acid and silver nitrate. Precipitated silver halides and silver sulphide are separated and weighed.

Carnallite. A mineral consisting of a mixed chloride of postassium and magnesium, $KCI.MgCI_2.6H_2O$, used as a source of postassium fertilizers.

Carnivore. Term applied to any animal whose diet consists primarily of animal matter. In animal systematics, it refers to members of the mammalian order Carnivora, which contains aquatic and terrestrial species, including the dog, cat, seal, and other families. The term *herbivore* refers to animals whose diets consist mainly of plant matter; *omnivore* refers to animals that eat animal and plant matter.

Carnotite. A narutally occurring vanadate of uranium and potassium, A $K_2(UO_2)(VO_4)_2 \cdot 3H_2O$. It is an important ore of uranium and vanadium.

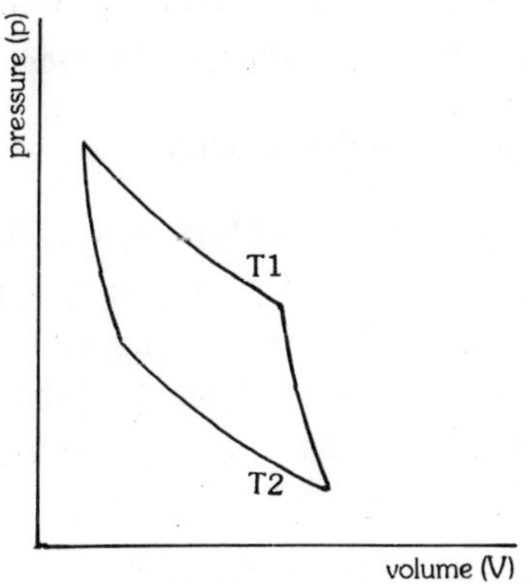

Fig. C-4. Carnot's cycle.

Carnot, (Nicholas Leonard) Sadi. 1796-1832, French physicist, son of Lazare Carnot. He studied the relation between heat and mechanical energy and devised an ideal engine whose series of operations showed that even under ideal conditions a heat engine must reject some heat energy instead of converting it all into mechanical energy. This illustrates the second law of Thermodynamics, formulated later.

Carnot's cycle. A reversible cycle of four operation applied to the gas in a heat engine. The successive stages are adiabatic compression, isothermal expansion, adiabatic expansion, and isothermal compression, thus restoring the original pressure, volume, and temperature. The Carnot cycle is the cycle of operations occurring in a perfect heat engine. [After Sadi Nicolas Leonard Carnot (1796-1832), French scientist.]

Carnot's theorem. The principal that the efficiency of a reversible heat engine depends only on its temperature range, not on the working substance.

Carob. Evergreen tree *(Ceratonia siliqua)* of the PULSE family, native to the Mediterranean but cultivated in other warm areas. Its large, red ponds have been a food source since prehistoric times, and the pods and their extracts have many common names, e.g., St. John's bread and locust bean gum. Carob is used as a food stabilizer and a caffeine-free chocolate substitute; carob pods are used as livestock feed.

Caro's acid. *See* peroxosulphuric acid.

Carp. Freshwater, bottom-feeding FISH *(Cyprinus carpio,* largest of the Minnow family, native to Asia but now found throughout Europe and America. Carp may be up to 3 ft 90 cm) and 25 lb (11.3kg). They have four "whiskers" (barbels) around the mouth, and are usually dark greenish or brown with red on the fines.

Carpet or **rug.** Thick fabric, originally woolen but now often synthetic, most commonly used as a floor covering. Of ancient origin, the art of carpet weaving reached its height in the handloomed Oriental carpets from 16th cent. Persia, Turkey, and central Asia. European carpets, with many outstanding examples dating from the 17th cent. in France and the 18th cent. in England were handmade until the introduction of the power loom by Erastus Bigelow in 1841. Although handmade rugs are still produced, contemporary carpet manufacturing has become a highly mechanized industry. Classifications of both antique and modern carpets include Oriental, European handwoven, velvet, chenille, rag, hooked straw, and fiber.

Carrier. 1. The entity, either an electron or positive hole, that carries the electric charge in a conductor or semiconductor.

2. The normal form of a chemical compound to which the

radioactive form is added in order to introduce it into a system for tracer studies.

Carrier gas. The gas used to carry the sample through the column in gasliquid chromatography.

Carrier wave. The radio wave of constant amplitude and frequency that is modulated *(see* modulation) by the signal in radio transmission.

Carrot. Common name for some members of Umbelliferae (also called parsley family), a family of mainly perennial or biennial herbs of north temperate areas. Most are typified by aromatic foliage, a dry fruit that splits when mature, and an umbellate inflorescence (in which the floret stems of the flattened flower cluster arise from the same point, like an umbrella). The seeds and leaves of many of these herbs are used for seasoning or as greens, e.g., Anise, Caraway, Coriander, Cumin, Dill, Fennel, and Parsley. The carrot, Celery, and Parsnip are commercially important vegetables. The common carrot (*Daucus carota sativa)* is a root crop, probably derived from the wild carrot (or Queen Anne's Lace). Carrots are rich in carotene (vitamin A), especially when cooked; in antiquity they were used medicinally. Some types, e.g., button snakeroot and sweet cicely, are used as aromatic ornamentals. A few members of the family, e.g., Poison Hemlock, produce lethal poisons.

Cartesian coordinates. Coordinates used to locate the position of a point with respect to intersecting lines *(axes),* each coordinate being the distance from one axis measured parallel to the others. In a plane, a horizontal x axis and *a* vertical y axis is used, the point *(a,b)* being a distance a (the *abscissa)* along the direction of the x axis and b (the *ordinate)* along the direction of the y axis. Usually the axes are perpendicular *(rectangular coordinates).* In three dimensions a z axis is also used. [After Rene Descartes (1596-1650), French philosopher and mathematician.]

Cartesian coordinates. System for representing the relative positions of point in a plane or in space. In a plane, the point *p* is specified by the pair of numbers *(x,y)* representing the distances of the point from two intersecting straight lines, referred to as the *x*-axis and the *y*-axis. The point of intersection of these axes, which are called the coordinate axes, is known as the origin. If the axes are perpendicular, as is commonly the case, the coordinate systems is called rectangular; otherwise it is oblique. In either type the x-coordinate, or abscissa, of *p* is measured along a line parallel to the *x*-axis, and the *y*-coordinate, or ordinate, along a line parallel to the *y*-axis. A point in space may be similarly specified by the triple of numbers *(x-y,z)* representing the distances from three planes determined by three intersecting straight lines not all in the same plane. Named for the French philosopher and scientist Rene Descartes, Cartesian coordinates allow certain question in geometry to be transformed into questions about numbers and resolved by means of Analytic Geometry.

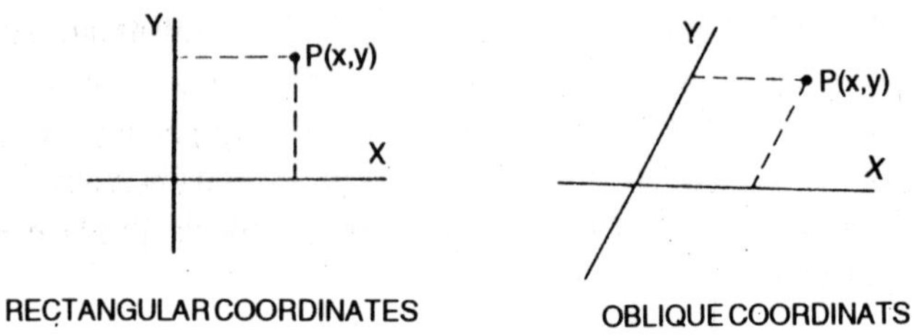

Fig. C-5. Cartesian coordinates.

Cascade. A process arranged in a series of consecutive stages so that the product of one stage is used as the starting material for the next. In the separation of isotopes by diffusion, the enriched material from one stage is further enriched in the next, and so on.

Cascade liquefier. A system for liquefying a gas by cooling it below its critical temperature and applying a pressure. A gas with a high critical temperature is first liquefied by pressure, then allowed to evaporate under reduced pressure thus lowering its temperature. This gas cools a second gas below its critical temperature, so that it can also be liquefied and evaporated to reduce the temperature still further. In this way the temperature is reduced in stages. If air is used liquid oxygen can be produced. The process cannot be used for hydrogen or helium because their critical temperatures are too low.

Casehardening. In metallurgy, a process to harden steel by increasing the percentage of carbon at its surface. This is done by packing the Steel in charcoal and then heating it; by heating it in a furnace with a hydrocarbon gas atmosphere; or by heating it in a molten-salt bath containing potassium and sodium cyanides.

Casein. A group of proteins found especially in milk is treated with acid, casein separates as an insoluble white curd; it is used in cheese, adhesives, and water paints. Rennet-curded casein is used to make cheese and a plastic from which imitation gemstones and other objects are made.

Cashew. Tropical American tree *(Anacardium occidentale)* of the Sumac family, valued chiefly for its kidney-shaped nut, whose sweet, oily kernel is used for food and yields an oil used in cooking. The nut grows at the end of a red, white, or yellow pear-shaped fleshy stalk, or cashew apple, which is also eaten or pressed to extract the juice, which may be fermented to make wine. The acrid sap of the cashew tree is used to make a varnish that protects woodwork from insects.

Cassegrainian telescope. A type of astronomical telescope in which the light is reflected by large concave mirror onto a smaller convex mirror, and from there through a hole in the main mirror into the eyepiece (*see* illustration). [After N. Cassegrain. 17th century French astronomer.]

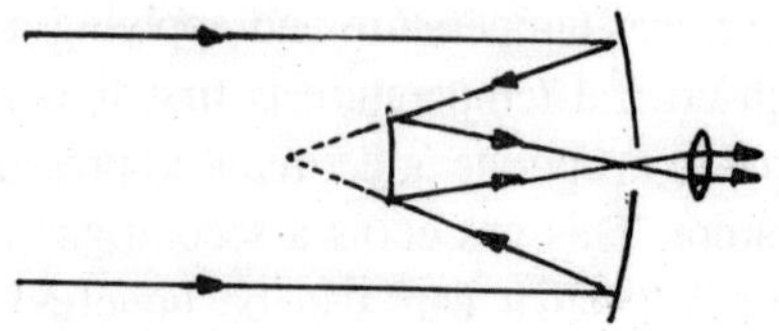

Fig. C-6. Cessegrainian telescope.

Cassiopeium. *See* lutetium.

Cassiterite. A black, brown, or white mineral consisting of tin IV oxide, SnO_2. It is the principal ore of tin.

Cassowary. Flightless forest Birds of Australia and the Malay Archipelago, smaller than the Ostrich. The plumage is dark and glossy, with a brilliantly-colored head and neck. Nocturnal mostly herbivorous, they are fast runners. Cassowaries are notoriously vicious, and have killed people with their sharp, spikelike toenails.

Casting or **founding.** The shaping of Metal by melting and pouring into a mold. Most castings are made in sand molds. Sand mixed with a binder to hold it together, is pressed around a wooden pattern that leaves a cavity in the sand. Molten metal is poured into the cavity and allowed to solidify. *Investment casting* is used for small complex shapes. Wax or plastic replicas of the parts are covered with sand in a box. When the whole mold is heated, the replica melts, leaving behind a cavity into which the metal is poured. *Die casting,* in which molten metal is forced under pressure into metal molds, is used to make large numbers of small, precise parts with metals of low melting points.

Cast iron A brittle iron-carbon alloy containing between 2% and 4.5% of carbon and small amounts of other elements such as sulpher, phosphorus, and silicon. The carbon is present as cementite or graphite.

Cast iron. From of Iron made from pig iron remelted in a small cupola furnace and poured into molds to make castings. It usually contains from 2% to 6% carbon, and scrap iron or steel is often added to vary the composition. Cast iron is used extensively to make machinc parts, engine cylinder blocks, stoves, pipes, radiators, and many other products.

Castor oil. A pale yellow viscous oil extracted form the seeds of the castor plant. It is used as a laxative and is also a raw material for the manufacture of resins, plastics, and lubricants.

Cat. Carnivorous Mammal of the family Felidae, including the domestic cat *(Fields catus),* the great cats (e.g., Lion, Tiger, Leopard, and Cheetah), and the similar wild cats (e.g., Lynx and Ocelot). Highly adapted for hunting and devouring their prey, cats have short muzzles large eyes, sensitive whisker, and sharp claws and teeth. Most have long tails and all have a flexible musculo-skeletal system. Lions and cheetahs live in groups called prides, but most cats are solitary. Cats have been domesticated since prehistoric time and were often the objects of veneration or superstitious fear. Domestic cats vary in size, with males usually weighing 9-14 lb (4.1-6.4kg) and females 6-10lb (2.2-4.5kg). They have cats of various length and many colors (black, white and shades of red, yellow, brown and grey) in a variety of patterns. Besides the common house cat, the species *F. catus* includes many recognised breeds with characteristics maintained by selective mating. Most breeds are short-hair. *Abyssinian cats* have ruddy brown cats with ticking (marking on each hair) of darker brown or black. *Burmese cats* are small and muscular with medium-to dark-brown coats; the *Manx cat* is a variously colored, tailless breed with hind legs longer than its forelegs, elevating the rump. *Siamese cats* have almond-shaped blue eyes and white, cream, or fawn coats with brown or grey areas (points) on the feet, tail, ears, and head. The *Russian Blue cat,* with green eyes and blue-grey coat, is distinguished by two layers of short, thick fur, and the *Rex cat,* the only curly haired cat, has a woolly coat that may be any color.

Persian cats are stocky and long-haired and may be variety of colors; the other long-haired breed, the *Himalayan cat* (a persian-Siamese cross), has Siamese coloring.

Catalysis. The process by which the rate of a chemical reaction is increased by the presence of another substance, the *catalyst*, that does not appear in the stoichiometric equation for the reaction. The catalyst only affects the rate of attainment of equilibrium, not the position of equilibrium, and it follows that the rates of the forward and reverse reactions must be affected equally. Thus dehydrogenation catalysts, such as platinum and nickel, are also good hydrogenation catalysts.

A distinction is generally made between *homogeneous catalysis*, in which the entire reaction occurs in a single phase, and *heterogeneous catalysis*, in which the catalyst and the reaction are in different phases. In the latter case the catalyst is usually solid and the reactants are gases.

Catalyst. Substance that causes a change in the rate of a chemical reaction without itself being consumed by the reaction. Catalysts, which walk by changing a reaction's activation energy, or minimum energy needed for the reaction to occur, are used in numerous industrial processes. Substance that increase the reaction rate are called positive catalysts, or simply catalyst, whereas substance that decrease the reaction rate are called negative catalysts, or inhibitors. The presence of a small amount of an acid or base may catalyze some reactions. Finally, divided metals (e.g., platinum, copper, iron, palladium, rhodium) or metal oxides (e.g., silicon dioxide, vanadium oxide) may also serve as catalysts. Biological catalysts are called Enzymes.

Cataract. Opacity of the lens of the Eye, causing impairment of vision. Most commonly caused by aging, cataracts may also be congenital or result from eye inflmmations or certain diseases, such as Diabetes. Cataracts are treated by surgical removal of the lens, implantation of an artificial lens, and the use of corrective lenses.

Catastrophism. Originally, the outdated geological doctrine that the physical features of the earth's surface, e.g., mountains and valleys, were formed during violent worldwide cataclysms, e.g., earthquakes and floods. This theory, easily correlated with religious beliefs, was systematized by Georges Cuvier, who argued that all living things were destoryed and replaced with wholly different forms during these cataclysmic events. In the 18th cent. The doctrine was attacked by James Hutton, who advanced the doctrine of Uniformitarianism. Geologists now use the term *catastrophe* to refer to a more localized, relatively milder event in earth history.

Catchment area or ***drainage basin.*** Area drained by stream or other water body. The amount of water reaching a River, reservoir, or Lake from its catchment area depends on the size of the area, amount of precipitation, and loss through evaporation and absorption.

Catechol (pyrocatechol, 1,2-dihydroxybenzene). A colourless crystalline substance, $C_6H_4(OH)_2$, used as a developer in photography. M.pt. 105°C; b.pt. 240°C; r.d. 1.4.

Catecholamine. Any of several structurally related compounds occurring naturally in the body that help regulate the sympathetic Nervous System. They include Epinephrine (or adrenaline), norepinephrine, and dopamine-substances that prepare the body to meet emergencies such as cold, fatigue, and shock. Epinephrine and isoproterenol, a synthetic catecholamine, are used as drugs to treat diseases such as Emphysema, Bronchitis, and Asthma.

Catenary. The curve in which a uniform rope would hang under gravity when suspended at each end. It has the equation $y = a(e^{x/a} + e^{-x/a})/2$.

Catenation. The formation of chains of atoms in molecules.

Catfish. Freshwater Fish (suborder Nematognathi) with barbels, or whiskers, around a broad mouth, fleshy, rayless posterior fins, scaleless skin, and sharp defensive spines on the pectoral and dorsal fines. One species, the electric catfish, discharges electricity (see Electric Fish). Catfish range in size from a few inches to 13 ft (3.9 m) or more. Omnivorous feeders and scavengers, they are important food fish and are raised on fish farms in the S. U.S.

Cathetometer. A device for measuring heights and lengths at a distance, consisting of a small telescope fitted with cross-wires and mounted so that it can move along a graduated vertical scale.

Cathode. An electrode with a negative electric charge, as in a cell, thermionic valve, rectifier, etc. *Compare* anode.

Cathode-ray oscilloscope (CRO). An instrument for displaying changing electrical signals on a cathode-ray tube. The signal to be investigated is applied, via amplifier to the vertical electrostatic defecting plates of the tube. The spot on the screen is deflected by an amount proportional to the strength of the signal. The beam is also deflected horizontally by the time base, which moves the spot horizontally across the screen at a uniform rate, returning it quickly to the starting point and repeating the sweep. The trace on the screen is graph of signal against time.

Cathode rays. Streams of electrons emitted from the cathode of a gas-discharge tube, vacuum tube etc.

Cathode-ray tube (CRT). An electronic tube for producing a visual display form electric signals, as used in television sets, radar screens, and cathode-ray oscilloscopes. It consists of an electron gun focusing a beam of electrons onto a fluorescent screen. The beam can be deflected, from side to side or up and down, by electrostatic or magnetic fields. Electrostatic deflection is effected by potentials applied to two pairs of plates mounted inside the tube on either side of the axis. In electromagnetic deflection wire coils are mounted outside the tube and the deflecting fields is produced by a current in these coils.

Cathode-ray tube. Electron Tube in which electrons are accelerated by high-voltage anodes, formed into a beam by focusing Electrodes, and projected toward a phosphorescent screen that forms the face of the tube. The electron beam leaves a bright spot wherever it strikes the screen. To form the screen display, or image, the electron beam is deflected in the vertical and horizontal directions either by the electrostatic effect of electrodes within the tube or by magnetic fields produced by coils located around the neck of the tube. Some cathode-ray tubes, made for Computer Terminals, Oscilloscopes, and Television receivers, can produce multiple beams of electrons and have phosphor screens that are capable of displaying more than color.

Cation (positive ion). An ion with a positive charge. Such ions are attracted to the cathode during electrolysis. *Compare* anion.

Catnip or **catmint.** Strong-scented perennial herb *(Nepeta cataria)* of the Mint family, native to Europe and Asia. Catnip, best known for its stimulated effect on cats, is also used to make a home-remedy tea.

CAT scan or **computerized axial tomography.** X-ray technique allowing safe, painless, and rapid diagnosis in previously inaccessible areas of the body. An X-ray tube, rotating around a specific ares of the body, delivers an appropriate amount of X-radiation for the tissue being studied and takes picture of that part of the internal anatomy from different angles. A computer then assists in forming a composite, readable image. CAT scanning has revloutionized medicine, especially neurology, by facilitating the diagnosis of brain and spinal cord disorders, Cancer, and other conditions.

Cattail or **reed mace.** Perennial herb (genus *typha),* found in open marshes. Cattails, or clubrushes, have long, narrow leaves and one tall stem with tiny male flowers above the female flowers. Pollinated female flowers from the familiar cylindrical spike of fuzzy fruits; the male flowers drop off. The starchy rootstocks are edible.

Cattle. Ruminant Mammals (genes *Bos),* especially a domesticated species. The term *oxen* is often used synonymously, although strictly speaking *ox* refers to mature, castrated male used for draft purposes. Western, or European, domestic cattle *(B. taurus)* are thought to be descended from the aurochs, a large wild ox domesticated in the Stone Age. The Zebu is the domesticated species of Asia and Africa. A grown male is called a bull; a grown female, a cow; an infant, a calf; a female that has not given birth, a heifer; and a young castrated male, a steer. Cattle are bred for beef and Dairying and for use as draft animals. Beef breeds include the Angus, Charolais, and Hereford; dairy breeds include Brown Swiss, Guernsey, Holstein-Friesian, and Jersey; dualpurpose (beef-dairy) breeds include the Red Poll and shorthorn.

Cauchy, Augustin Louis, Baron (koshe). 1789-1857, prolific French mathematician who did influential work in every branch of Mathematics. In calculus, he invented the notion of continuity, gave the first adequate definition of the definite integral as limit of sums, and defined improper integrals. Cauchy provided the first comprehensive theory of complex functions, published the first comprehensive treatise on determinants, and founded the mathematical theory of elasticity.

Cauliflower. Variety of Cabbage (var. *botrytis),* with an edible head of condensed flowers and stems. Another cultivation of the same variety is broccoli. Both have been grown since Roman times.

Caustic. 1. A curve or surface that is the envelope of light rays reflected or refracted at a curved surface. Such rays are not focused to one point. They are all tangential to the cusp-shaped caustic curve (see spherical aberration).

2. Denoting a corrosive alkaline substance.

Caustic potash. *See* potassium hydroxide.

Caustic soda. *See* sodium hydroxide.

Cave. A hollow in earth or Rock. Caves are formed by the chemical and mechanical action of water on soluble rock, by volcanic activity (the formation of large gas pockets in lava or the melting of ice under lava), and by earthquakes. Limestone formations, due to their solubility, almost invariably have caves, some notable for their Stalactites and Stalagmites. Among famous U.S. caves are the Carlsbad Caverns in New Mexico.

Cavendish's experiment. An experiment for determining the gravitational constant and demonstrating the inverse-square law of gravitation. A long thin beam with a small leave sphere at each end was suspended by a wire. Large lead spheres were placed close to the smaller spheres, first to one side of each sphere and then to the other. Using the system as a torsion balance, it was possible to calculate to *G* from the change in angular deflection [Performed in 1798 by Henry Cavendish (1731-1810), English Scientist.]

Cavendish, Henry. 1731-1810, English physicist and chemist; b. France. He determined the specific heats for numerous substances (although these heat constants were not recognized until later), did research on the composition of water and air, and studied the properties of a gas that he isolated and described as "inflammable air" (later named Hydrogen). In a now-famous experiment (1798), he determined the value of the proportionality constant in Newton's law of Gravitation.

Cavitation. The formation of small vapour-filled bubbles in a liquid regions in which it has a high velocity. Cavitation occurs around propellers, in pumps, in similar devices. The cavities formed in the low pressure region are carried into a region of higher pressure where they collapse, causing pitting of the surface.

Cayley, Sir George. 1773-1857, English scientist, recognised as the founder of Aerodynamics. In his studies on the principles of flight, he experimented with wing design, distinguished between left and drag; and formulated the concepts of vertical tail surface, steering rudders, rear elevators, and air screws. Although powered flight was not possible in his time, he was able to calculate the power required for various speeds and loads.

Cedar. Common name for a number of mostly coniferous evergreen trees. The true cedars (genes *Cedrus)* of the Pine family, are all native to Old World; some, e.g., the cedar of Lebanon *(C. libani)* and the fragrant deodar cedar *(C. deodara)*, are cultivated elsewhere. In North America, the name cedar refers to the Juniper (red cedar). Arborvitae (white cedar), and other conifers of the Cypress family. Several tropical American trees (genes *Cedrela)* of the Mahogany family are also called cedars.

Celery. Biennial plant *(Apium graveolens)* of the Carrot family, widely distributed and cultivated in north temperate areas. Once used as a medicine and flavoring, it is now used chiefly as a food, especially in soups and salads; the seeds are still used for seasoning.

Celestial equator. The great circle on the celestial sphere parallel to the earth's equator and perpendicular to the great circle passing through the celestial poles.

Celestial mechanics. The branch of astronomy concerned with the relative motion of celestial objects under the influence of gravitational forces.

Celestial mechanics. The study of the motion of astronomical bodies as they move under the influence of their mutual Gravitation. The calculation of such motion is complicated because many separate forces are acting at once and all bodies are moving simultaneously. Celestial mechanics is based on Isaac Newton's laws of motion and theory universal gravitation. Only the problem of two isolated moving bodies mutually attracted by gravitation can be solved exactly. Because the sun is dominant influence in the solar system, an application of the two-body problem leads to the simple elliptical Orbits as described by Kepler's Laws, which give a close approximation of planetry motion. Problems that consider the additional effects, or perturbations, of other less dominant bodies (such as the other planets in the solar system) cannot be solved exactly except in a few special cases. Methods have been devised, however, to allow successive refinements of an approximate solution to be made to almost any degree of precision.

Celestial poles. The northern and southern points in the sky where the earth's axis would intersect the celestial sphere.

Celestial sphere. The sky thought of as the inner surface of a sphere of infinite radius, centred on the position of the observer or the centre of the earth, upon which celestial objects appear to be projected. The fixed position of the stars are specified with respect to the celestial equator and the celestial poles; the positions of celestial objects as they change relative to the observer are giving with reference to the celestial poles, the horizon, and the zenith. Other fixed coordinate system are based on the ecliptic and the plane of the Galaxy.

Celestine. A mineral, white, blue in colour, consisting of strontium sulphate, $SrSO_4$. It is source of strontium salts.

Cell. In biology, the unit of structure and function of which of plant and animals are composed. The cell is smallest unit in the leaving organism that is capable of carrying on the essential life processes of sustaining Metabolism for producing energy and reproducing. Many single-called organisms (e.g., Protozoa) perform all life functions. The cell is differentiated into the cytoplasm, the cell members, which surrounds it; and the nucleus which is contained within it. Plant cell also have a thickened cell wall, composed chiefly of Cellulose. Included in the cytoplasm are mitochondria, which produce energy; lysozymes, which digest; Golgi apparatuses, which synthesize, store and secrete substances; ribosomes, the sites of protein synthesis; and Chloroplasts (in green plants only), in which Photosynthesis occurs. The nucleus contains chromosomes, which store the information for the metabolic functions of the whole cell and pass on their information to daughter cells by replicating themselves exactly. In higher organisms groups of cells are differentiated into specialized tissues.

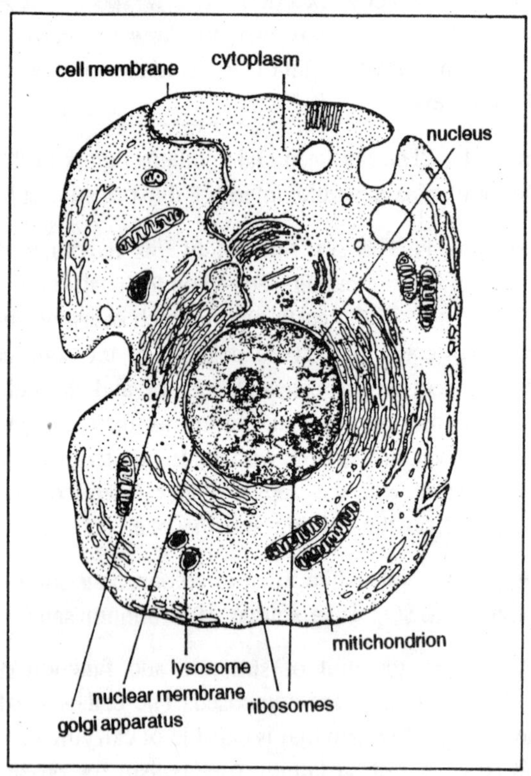

Fig. C-7. Animal cell.

Cell. In electricity, device that operates by converting chemical energy into electrical energy. A cell consists of two dissimilar substances, a positive Electrode and a negative electrode, and a third substance, the Electrolyte, that acts chemically on the electrodes. A group of cells connected together is celled a battery. The Electromotive Force, or voltage produced between the positive and negative electrode, depends on the chemical properties of the substance used but not on a size of the electrodes or the the amount of electrolyte. When the electrodes are connected externally by a piece of wire, electrons flow out of the negative electrode, though the wire, and into the positive electrode. There are several kinds of cells, differing in electrode material and electrolyte. The Leclanche cell has

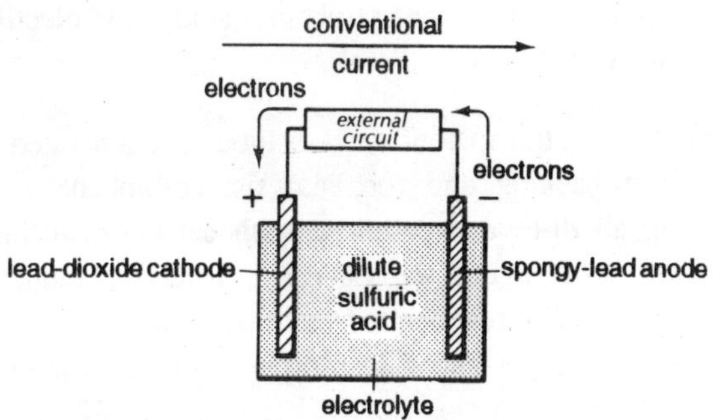

Fig. C-8. Cell, in electricity: At the lead-dioxide cathode, electrons from the circuit combine with lead dioxide and sulfuric acid to form lead sulfate and water. At the spongy-lead anode, lead reacts with sulfate ions to form lead sulfate and release electrons.

a zinc negative electrode, a carbon positive electrode, and an electrolyte consisting of ammonuium chloride solution. It is the basis of the common dry cell, so called because the electrolyte is in the form of a paste instead of a liquid. An alkaline dry cell, which can operate up to ten times longer than common dry cells, has a zinc negative electrode, a manganese dioxide positive electrode, and an electrolyte of potassium hydroxide. A mercury dry cell, with a zinc negative electrode, a mercuric oxide positive electrode, and a potassium hydroxide electrolyte, has a constant output voltage and may be stored for many years.

Cell. **1.** A device for converting chemical energy directly into electrical energy, consisting of two electrodes dipping into an electrolyte. Ions are produced of discharge and each electrode so that one acquires a positive potential and the other negative potential. When the electrodes are connected through a circuit, a current flow. A *primary cell* is one in which the electrical energy is produce by the reaction occurring in the cell. A *secondary cell* (or accumulator) is one used for storing electricity.

2. Any one various similar devices having electrodes and an electrolyte used for electrolysis, study of electrochemical processes, etc.

Cellophane. Thin, transparent sheet or tube of regenerated Cellulose. Used in packing and for Dialysis, cellophane is made by mixing alkali-treated cellulose with carbon disulfide to form viscose. After aging, the viscose is forced through a slit into dilute acid. The regenerated cellulose that results has a lower molecular weight and a less orderly structure than cellulose.

Celluloid. Transparent, colorless synthetic Plastic made by treating Cellulose nitrate with Camphor and alcohol. The first important synthetic plastic, celluloid was widely used as a substitute for more expensive substances, such as ivory, amber, and tortoiseshell. It is highly flammable and has been largely superseded by newer plastics.

Cellulose. A carbohydrate of high molecular weight that is the chief constituent of Cell walls of plants. Raw cotton is 91% cellulose. Other important natural sources are flax, hemp, jute, straw, and wood. Cellulose has been used to make Paper since the 2d cent. A.D. Cellulose derivatives include

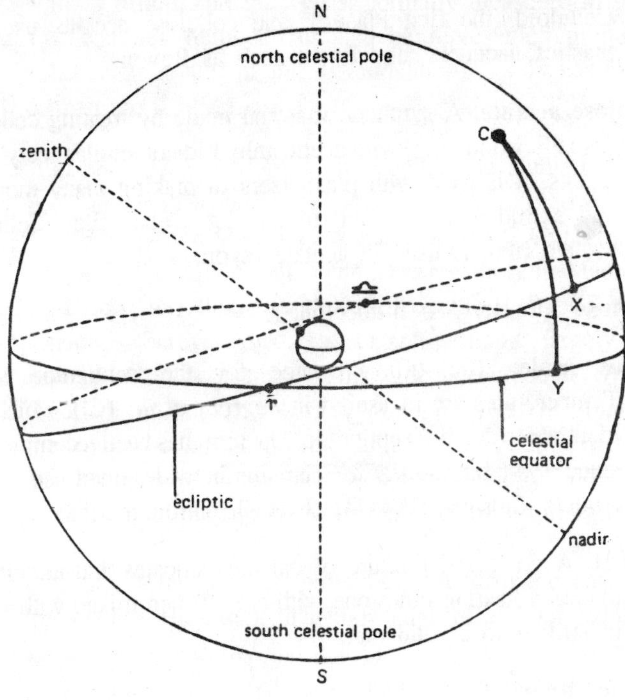

♈ first point of Aries; vernal equinox

♎ first point of Libra; autumnal equinox

C celestial object

♈Y right ascension of C (in hours anticlockwise from ♈)

♈X celestial longitude of C (in degrees anticlockwise from ♈)

YC declination of C

XC celestial latitude of C

Fig. C-9. Celestial coordinates

guncotton (fully nitrated cellulose), used for Explosives; Celluloid (the first Plastic); and cellulose acetate, used for plastics, lacquers, and fibers such as Rayon.

Cellulose acetate. A synthetic material made by treating cellulose (usually wood pulp) with acetic anhydride or similar acetylating agents. It is used with plasticizers in making many moulded and extruded articles and in making acetate film. Cellulose acetate fibre is used in acetate rayon.

Cellulose nitrate. *See* nitrocellulose.

Celsius scale. The official name for the centigrade scale. Temperatures are measured in *degrees elsius* (C°), which are identical to degrees centigrade. The term has been recommended since 1948 but *centigrade* remains in widespread use. [After Anders Celsius, 1701-44, Swedish astronomer.]

Cement. A powdered mixture of calcium silicates and aluminates made by heating limestone with clay. When mixed with water it hardens to a solid mass.

Cement, hydraulic. Building material typically made by heating a mixture of limestone and clay until it almost fuses and then grinding it to a fine powder. Once it is mixed with water, cement will harden even if immersed in water. The most common cement, portland cement is made by mixing and then heating substance containing lime, silica, alumina, and iron oxide, with gypsum added during the grinding process. Quick-setting aluminous cement is made from limestone and bauxite. see also Concrete.

Cementite. Iron carbide, Fe_3C, present in cast iron and steel.

Centre of mass. The point at which all the mass of a body of group of objects is considered to be concentrated. This point is often the same as the centre gravity (the point at which all the weight of a body may be considered to be concentrated). The motion of several colliding elementary particle, or of the earth and moon around the sun, are sometimes analyzed from the point of view of the centre of the mass of the entire system of objects.

Centi-. Symbol: c. Prefix indicating one hundredth (10^{-2} e.g. 1 centimeter (cm) = 10^{-2}m).

Centigrade scale. A temperature scale in which the melting point of ice is 0° and the boiling point of water is 100°. Temprature is measured in *degrees centigrade* (°C). Since 1984 the recommended name has been the *Celsius scale.*

Centimetric waves. Microwaves with wavelengths in the range 1-10 centimetres.

Centipede. Invertebrate animal of the class Chilopoda, an Arthropode. The flattened body is divided into a head and truck comprised of segments (somites). The average number of legs is 35 pairs, one pair per segment except for the first one and last two. The appendages of the first segment are modified into claws equipped with poison glands, used to capture prey. The largest tropical species may reach 12 in. (30 cm) length; temperate species are usually about 1 in. (2.5 cm) long.

Central force. A force on a moving body that is always directed towards a fixed point. A planet, for example, experiences a central force directed towards the sun.

Central processing unit (CPU). The part of a digital computer in which operations are performed on the data.

Centre of gravity. A fixed point through which the weight of a body can be considered to act, irrespective of the body, position.

Centre of mass. A fixed point at which the mass of body can be considered to act. In a uniform gravitational field the body's centre of mass coincides with its centre of gravity.

Centrifugal force. A force that appears to act radially outwards on a body that moves in a curved path. It is equal and opposite to the centripetal force. A body with mass m moving in a circular path of radius r with velocity v suffers a centrifugal force mv^2/r.

Centrifugal separation. A method of separating isotopes by using a large centrifuge. The technique can be applied to both liquid and gaseous samples. Under the action of the centrifugal force, radial concentration gradients are set up in which the lighter isotope tends to concentrate nearer the axis.

Centrifuge. Device using centrifugal force to separate substance density, e.g., cream from milk. Substances are placed in containers that are spun at speeds high enough to cause the heavier element to sink. The first successful centrifuge was built in 1883 by the Swedish engineer Carl G.P. de Laval.

Centrifuge. An apparatus for separating suspended particles in a liquid by centrifugal force. It consists of a rapidly rotating bar or wheel with tubes attached containing the suspension, free to pivot outwards at high seed.

Centripetal force. A force acting radially inwards constraining a body to move in a curved path. It is equal and opposite to the centrifugal force.

Centripetal force and centrifugal force. Action-reaction force pair associated with circular motion. Centripetal ("centre-seeking") force is the constant inward force necessary to maintain circular motion the centripetal force F = acting on a body of mass m is given by the equation $F = mv^2/r$, where v is its velocity and r is the radius of its path. The centripetal force, the action, is balanced by a reaction force, the centrifugal ("centre-fleeing") force, which acts not on the circling object what on the source of centripetal force, usually located at the circle's center. The forces are equal in magnitude and opposite in direction.

Centroid. The point associated with a geometric figure or solid at which the centre of mass would occur if the geometric object were made of homogeneous material. The centroid of a plane figure, for example, is the centre of mass of a thin sheet of material with the same shape and dimensions as the figure.

Cephalopod. Free-swimming Mollusk (class Cephalopoda) with a long body with a ring of sucker-bearing arms (tentacles) encircling the mouth. The head is large, with prominent eyes. Only the Nautilus has an external shell-in the Squid and Cuttlefish the shell is internal and reudced, and in the Octopus it is completely absent. Most cephalopods are aggressive carnivores that move by means of a king jet propulsion.

Cepheid variables. Small class of Variable Stars that brighten and dim in regular periods ranging from 1 to 50 days. The periods of Cepheids, which are yellow supergiant stars that swell and contract in size, depend on their intrinsic brightness, absolute Magnitude, in known way: the brighter the star, the greater its period. By comparing the Cepheid's absolute magnitude to its apparent magnitude, one can determine its distance. The period luminosity relation of Cepheids make them invaluable in estimating interstellar and intergalactic distances.

Ceramic. Any of various hard high-melting nonmetallic inorganic materials, including pottery, porcelain, enamels, and refractory substances. Many ceramics are metal silicates and aluminosilicates, obtained by fusing clay. Other examples include metal oxides, brides, carbides, and nitrides and such compounds as ferrites and titanates. *See also* cermet.

Cerebral palsy. Disorder in which muscular control and coordination are impaired, and speech and hearing problems and mental retardation may occur. It is most commonly caused by brain damage occurring before during but, varies greatly in severity. Treatment involves physical, occupational, and speech therapy;

braces; orthopedic surgery; and drugs to reduce muscle stiffness and spasticity.

Cerenkov radiation. Bluish light emitted when charged particles move through a medium at a velocity greater than the velocity of light in that medium. The effect is used as the basis for a particle counter *(Cerenkov counter)* by detecting the light with a photomultiplier. [After Pavel Alekseyevic Cerenkov (b. 1904), Soviet physicist.]

Cerium. Symbol: Ce. A grey malleable ductile metallic element belonging to the lanthanide series. It forms salts with a valency of 3 *(ceric)* or 4 *(cerous)*. A.N. 55; A.W. 132.9055; m.pt. 28.4°C; b.pt. 678.4°C; r.d. 1.87.

Cermet. Any of a class of materials made by sintering or bounding together powdered ceramic and metal. Cermets are hard and resistant to corrosion and high temperatures.

Cesarean section. Delivery of an infant by surgical removal from the uterus through an abdominal incision. It usually performed when childbirth is considered hazardous, as when the mothers pelvis is too narrow, the fetus is in an abnormal position, or the mother is older than usual at the time of her delivery. Subsequent deliveries are usually also by cesarean section. The name comes from the legend that Julius Caesar was born in this manner.

Cesium (Cs). Metallic element, discovered by spectroscopy in 1860 by Robert Bunsen and Gustav Kirchhoff. Ductile, soft as wax, and silver-white, it is the most alkaline element (see Alkali Metals) and the most reactive metal. Cesium metal is used in photoelectric cells and various optical instruments; cesium compounds, in glass and ceramic production. The cesium-137 radioactive isotope is used to treat cancer.

Cetane rating. A number measuring the ignition qualities of diesel fuel. It is the volume percentage of cetane (hexadecane;

$C_{16}H_{34}$) in a mixture of cetane and a-methylnaphthalene ($C_{10}H_7CH_3$) with the same ignition qualities as the fuel tested. A similar measure, the octane rating, is used for petrol.

CGS units. A system of units based on the centimetre, the gram, and the second. The system includes the dyne as the unit of force and the erg as the unit of energy. The calorie is a unit of heat energy.

Two types of electrical unit are used in theCGS system: electrostatic E_o units in which the electric constant, is unity and electomagnetic units in which the magnetic constant, μ_o is unity. The constants are related by $1/(E_o\mu_o)^{1/2} = c$, where c is the velocity of light.

In practice, *Gaussian units* have usually been used. In these the units of electrical quantities *(Q, V, I, E, etc.)*. are the same as the electrostatic units and the units of magnetic quantitics *(H,B, Ø,* etc.) are the same as the electromagnetic units. Equations including quantities from both groups contain a factor C.

Chadwick, Sir James. 1891-1974, English physicist. He worked on radioactivity under Ernest Rutherford and was assistant director (1923-35) of radioactive research at Cavendish Laboratory, Cambridge. For his discovery of the Neutron he received the 1935 Noble Prize in physics.

Chain reaction. Any process in which a product of one reaction initiates a further reaction, giving more product, which in turn causes further reaction, and so on.

In a *nuclear chain reaction* a series of fission reactions occur. A uranium-235 nucleus, for example, may capture a neutron and undergo fission. The fragments include two or three further neutrons. Under favourable conditions these may cause further fissions, leading to even more neutrons. The reaction becomes self-sustaining with the number of free neutrons increasing—it is said to be *supercritical.* On the other hand, the neutrons may not induce sufficient fission to cause a self-sustaining process-the reaction is *subcritical.*

In a *chemical chain reaction* the reaction is maintained by the product of one stage taking part in the next. Chain reactions often occur when free radicals are produced: a free radical may attack a molecule, removing a hydrogen atom and leaving another radical, which in turn attacks another molecule, etc.

Chain, Ernst Boris. 1906-79, English biochemist; b. Germany. For their work in isolating and purifying penicilline, he and Sir Howard Florey shared with Sir Alexander Flaming the 1945 Noble Prize in physiology or medicine. Chain held various teaching and research positions in Berlin, Rome and London and at Oxford and Cambridge.

Chair configuration. *See* configuration.

Chalcogens. The elements oxygen, sulphur, selenium, tellurium, and polonium, belonging to group VIA of the periodic table.

Chalcopyrite (copper pyrites). A yellow lustrous naturally occurring mixed sulphide of iron and copper, $(Cu,Fe)S_2$, important as the main ore of copper.

Chalcopyrite or **copper pyrites.** Copper and iron sulfide mineral ($CuFeS_2$), brass yellow and sometimes with an iridescent tarnish. It is found in crystal form but is most often massive. Occurring worldwide in igneous and metamorphic rocks, it is an important ore of Copper.

Chalk. Calcium carbonate mineral, similar in composition to limestone but softer. Chalk has been deposited throughout geologic time. The chief constituents are the shells of minute animals called foraminifera; however, the dominant component of the best -known formations, the Cretaceous chalks (e.g., the White Cliffs of Dover, England), are coccolith algae. Chalk is used to make putty, plaster, cement, quicklime, and blackboard chalk. Harder forms are used as buildings stones, and poor soils with high clay content are sweetened with chalk.

Chameleon. Small-to medium-sized Lizard of the family Chamaeleonidae, found sub-Sharanah Africa, with a few spaces in S.. Asia. Chameleons have laterally flattened bodies ornamented with crests, horns, or spines and bulging, independently rotating eyes. Their skin changes color in response to stimuli such as light, temprature, and emotion, not a response to background color. The so-called common chameleon *(Chamaeleo chamaeleon)* is found around the Mediterranean. The American chameleon, not a true chameleon, belongs to the Iguana family.

Chamois. Hollow-horned, hoofed Mammal *(Rupicapra rupicapra)* found in mountains of Europe and the E Mediterranean. About the size of a large Goat, it brown with a black tail and back strips; its horns are erect with terminal hooks. The skin was the original chamois leather, a name now also given to the skins of other animals.

Chamomile or **camomile.** Name for some herbs of the Composite family, especially the perennial English, or Roman, chamomile *(Anthemis nobilis)* and the annual German, or wild, chamomile *(Matricaria chamomilla)*. The former is the chamomile most used for ornament and for a tea, made from the dried flower heads containing a volatile oil. The oil form wild chamomile flowers is chiefly used as a hair rinse.

Champagne. Sparkling white wine, traditionally made from a mixture of grapes grown in the old French province of Champagne, the best is from the Marne valley. It was reputedly developed in the 17th cent. by a monk, Dom Perignon. The fermented and blended wine is bottled, then sweetened and allowed to ferment further. The carbonic acid left in the bottle after the final fermentation gives champagne its sparkle.

Change of state. The change of a substance of one state of matter to another. Changes of state include the processes of melting, boiling, freezing, condensation, and sublimation.

Chaparral. Type of plant community in which shrubs dominant, occurring usually in areas drier than forests and wetter than desert. The species of shrub vary in different areas. The chaparral in Colorado, E Utah, and N New Mexico is mostly deciduous, while that of S California, Nevada, and Arizona is primarily evergreen. Chaparral is well exemplified in the W and SW U.S., but similar growth is found in many other part of the world.

Characteristic. *See* logarithm.

Charcoal. An amorphous form of carbon made by the destructive distillation of wood or other organic matter. It is used as an absorbent.

Charcoal. Nonvolatile residue obtained when organic matter, usually wood, is heated in the absence of air. Largely pure Carbon, charcoal yields more heat volume than wood. Charcoal obtained from bones is called bone black or animal charcoal. Finelly divided charcoal, with its porous structure, efficiently filters the adsorption of gases and of solids from solution, Charcoal is used in sugar refining and water and air purification.

Charge. 1. Symbol Q. The property of certain elementary particles that enables them to attract or repel one another by electromagnetic interation. Charge is conventionally designated as positive or negative, such that like charges repel one another and unlike charges attract. The electrons has the smallest natural unit of negative charge with the proton having an equal amount of positive charge. A body of system has a negative charge if it contains more electrons than more protons and a positive charge if it contains more protons than electrons. Quantitatively, charge is taken to be the product of an electric current and the time for which it flows. If the current is not contain, the time integral is used. Charge is measured in coulombs. *See also* electrostatic units, electromagnetic units.

2. (of a capacitor) The electric charge on the positive plate of a parallel-plate capacitor.

3. The total amount of electricity stored in an accumulator. It is usually measured in ampere-hours.

Charge. In Electricity, property of matter that give rise to all electrical phenomena. The basic unit of charge, usually denoted by *e,* is that on the Proton or Electron; that on the proton is designated as positive (+e) and that on the electron is designated as negative (-e). All other charged Elementary Particles have charges equal to +e, e, or some whole number times one of these, with the possible exception of the Quark, a hypothetical particle whose charge could be $^{1}/_{3}$e or $^{2}/_{3}$e. Every charged particle is surrounded by an electric Field of force such that it attracts charge of opposite sign brought near it and repeals of any change like sign. The magnitude of this force is described by Coulonb's Law. This force is much stronger than the gravitational force between two particles and is responsible for holding protons and electrons together in Atoms and in chemical bonding. Any physical system containing equal numbers of positive and negative charges is neutral. Charge is a conserved quantity; the net electric charge in a closed physical system is constant. Although charge is conserved, it can be transferred from one body to another. Electric current is the flow of charge through a conductor.

Charge-transfer complex. A type of complex formed by partial electron transfer between two different molecules. Iodine forms such complexes with certain aromatic compounds.

Charm. A postulated property of certain elementary particles arising from the idea that a multiple of particles with SU3 symmetry (see unitary symmetry) may be subgroup of a larger group of particles with symmetry SU4. The octet of mesons, for example, may be a subgroup of multiple of fifteen particles. The grouping of particles using the SU3 symmetry group implies that elementary particles may be composed of three types of quark. If SU4 symmetry group do occur an extra quark is required having a charge 2/3, zero strangeness, and the extra property of charm. The additional seven mesons would consist of combinations of a charmed quark with the other types of quark. One of these containing the charmed quark and its antiquark, would have zero charm but the others would have a charm of +1 or -1. Similarly, charmed baryons may be expected to occur.

So for no charmed particles have been detected but the concept helps to explain certain particle-scattering processes and may also account for the existence and properties of the psiparticle.

Charles' law. The principal that a fixed quantity of gas at constant pressure has a volume that is inversely proportional to its temperature. If the thermodynamic temperature is used the law has the form $V = AT$, where A is a constant for all gases. The law strictly applies only two ideal gases: real gases tend to obey it at low pressures.

If the volume of a gas is known at 0°C, the volume increase or decrease by 1/273 of this volume for each degree rice of fall in temprature. This is sometimes given as a statement of Charles' law. The principal is also called *Gay-Lussac's law. See also* gas laws. [After Jaques Charles (1746-1823), French physicist.]

Cheese. Food known from ancient times, consisting of the curd of Milk separated from the whey. Although milk from various animals has been used for making cheese, today milk from cows, sheep, and goats is most common. In making cheese, casein, the chief milk protein, is coagulated by enzyme action, by lactic acid, or by both. The many kinds of cheeses depend for their distinctive qualities on the kind and condition of the milk, the processes used in their making, and the method and extent of curing. There are two main kinds of cheeses: hard cheeses, which improve with age, and soft cheeses, made for immediate consumption. Hard cheeses include cheddar (originally from England), Edam and Gouda (Holland), Emmental and Gruyere (Switzerland), and Parmesan and Provolone (Italy). Among the semisoft cheeses are Roquefort (France), American brick, and Meuenster. Soft cheese may be fresh (unripened), e.g., cream cottage cheese, or may be softened by microorganisms in ripening process that develops flavor, e.g., Camembert, Brie, and Limburger. Cheese is a valuable source of protein, fat, insoluble mineral, and, when made whole milk, vitamin A.

Cheetah. CAT *(Acinonyx jubatus)* found in Africa, SW Asia, and India. The swiftest four-footed animal, it runs down its prey at speeds of over 60mi (95 km) per hr, the only cat to hunt this way. Cheetahs have tawny coats with many rounds, black spots; the average adult weighs 100 lb (45 kg). Cheetahs are unique among cats in having nonretractile claws. Hunting has greatly reduced their numbers.

Chelate. An inorganic metal complex in which there is a closed ring of atoms, caused by attachment of a ligand to a metal atom at two points. An example is the complex ion formed between ethylene diamine and cupric ions, $[Cu(NH_2CH_2CH_2NH_2)_2]^{2+}$.

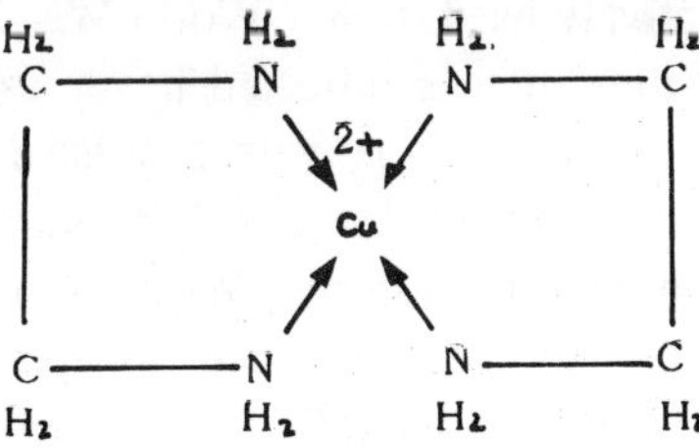

Fig. C-10. Copper chelate.

Chelating agent. A molecule, ion, or redical that can chelates.

Chelating agents are capable of coordinating to a metal atom at more than one point. If the ligand coordinates at two points it is said to be *bidentate:* ligand that coordinate at three points (forming two rings) are *tridentate*, etc. Chelating agents are often used for removing metal ions form solution or neutralizing their chemical activity. *See* sequestration.

Chemical combinations, laws of. Three laws describing the way in which elements combine together to form compounds. They are the laws of constant composition, multiple proportions, and reciprocal proportions. The third is derivable from the first two. *See also* Dalton's atomic theory.

Chemical equilibrium. The state of a reversible chemical reaction when its forward and reverse reactions occur at the same rates, so that there is no apparent change with time in the amount of products and reactants present. The relative amounts of products and reactants are then given by equilibrium constant.

Chemical equivalent. *See* equivalent weight.

Chemical reaction. A process in which one or more chemical substances change into the other substances by breaking or forming chemical bonds.

Chemical warfare. Employment in warfare of toxic substance to damage or killed plants, animals, or human beings. Poison Gas was effectively used during World War I, when chlorine gas and later mustard gas inflected heavy casualties on bath side. After the war major powers continued to develop and stockpile chemical agents for possible future use, but lethal types were not employed during World War II. Thousand of synthetic toxins and naturally occurring poisons have since been tested. Besides potentially lethal chemical that attack the skin, blood, the nervous system, or the respiratory system, there are also nonlethal incapacitating agents that cause temporary physical disability or mental effects such as confusion, fright, or stupor. Such agents, e.g., tear gas, may be used in riot control as well as warfare. Various forms of Herbicides and defoliants were used during the Vietnam War to destroy corops and clear away vegetation. In the early 1980s there were persistent reports that a lethal agents, popularly called, "yellow rain", was being used in Southeast Asia. Modern delivery system-artillery shells, grenades, missiles, and aircraft submarine spray systems- have increased the potential effectiveness of chemical warfare, as they have that of Biological Ware fare, in which microorganisms are the toxic agents.

Chemiluminescence. Luminescence occurring as a result of chemical reaction. Part of the energy of take up by exciting the electrons so that molecules of one of the products appear in excited state, from which they decay by emission ·or light. *Luminol,* for example, is a crystalline compound that emits blue light when oxidized with hydrogen peroxide.

Chemisorption. Adsorption in which the adsorbed substance is held by chemical bonds.

Chemistry. The branch of science concerned with the element and their componds. Chemistry involves the study of effects produced by the orbiting electrons of atoms, rather than by atomic nuclei. It is divides into inorganic, organic, and physical chemistry.

Chemistry. Branch of science concerned with the properties, compostion, and structure of substancses and the changes they udergo when they combine or react under spcified conditions. Inorganic chemistry deals mainly with components of mineral origin. Organic Chemistry was first defined as the study of substances produced by living organisms; it is now defined as the study of the compounds of Carbon. Physical Chemistry is concerned with the physical properties of materials; its subcategories are Electrochemistry; thermochemistry, the investigation of the changes in Energy and Entropy that occur during chamical reactions and phase transformation (see State of Matter); and chemical kinetics, which is concerned with the details of chemical reactions and of how equilibrium is reached between the products and reactants. Analytical chemistry is a colletion of techniques that allows exact laboratory determination of the chamical composition of a given sample of material.

Chemotherapy. Treatment of disease with chemicals or drugs; the term is now most often used to refer to treatment of Cancer. In current approaches, drugs are used in combination to create a synergistic effect, and doses are scheduled to attack rapidly proliferating cells, such as cancer cells, during their vulnerable phase. These drugs are administered to destroy or inhibit infecting organisms or malignant tissue, but are less toxic to normal tissue. Chemotherapy is highly individualized depending on the disease state, the action of the agents used, and the side effects in the patient.

Cherry. Name for various trees and shrubs (genus *prunus*) of the Rose family, and for their fruits. Botanically, the small red-to-black fruits are drupes, or stones fruits, closely related to the Peach Apricot, and Plum. Hundreds of varieties of sweet *(P. avium)* and sour *(P. cerasus)* cherries, believed to be native to Asia Minor, are Widely cultivated. Sour cherries are mostly self-fertile, and are hardier and more easily grown than sweet cherries, which must be cross-pollinated. The fruit is popular eaten raw, and in preserves, pies, ciders, and liqueurs. Species of *flowering cherry* are cultivated for their beautiful, usually double flowers; *cherry laurel* species are also cultivated as ornamentals, e.g., American cherry laurel, or mock orange *(p. caroliniana)*. The wood of the wild black cherry *(P. serotina)*, fine-grained and usually reddish in color, is prized for cabinetwork.

Chert. Cryptocrystalline variety of Quartz, commonly occurring in nodules. Flint-the dark variety of chert-was used by primitive peoples to make knives and spearheads, because, although it is very hard, it is easily shaped by flaking off the edges. It was long used with steel for lighting fires and later for setting off the powder in flintlock firearms.

Chestnut. Deciduous tree (genus *Castanea)* of the Beech family, with thin-shelled, sweet, edible nuts borne in bristly burs, widely distributed in the Northern Hemisphere. The common American chestnut *(C. dentata),* native to the E U.S., is nearly extinct from chestnut blight, a fungal disease, and edible chestnuts are now mostly imported from Italy. Some old chestnut forests still stand as dead, or "ghost", forests; the dead and fallen logs are the leading domestic source of tannin. Some American species are called chinquapin.

Chickadee. Small North American Bird of the Titmouse family. The lively black-capped chickadee *(Parus atricapillus),* with gray back and wings and light underparts, often swings upside down on branch tips, searching for insects.

Chicken. Chief poultry Bird, probably domesticated from a SW Asian jungle fowl. Chickens are raised commercially in highly mechanized indoor chambers and are classified by their uses. The Leghorn chicken is the main egg breed in the U.S., while a cross between the fast-growing Plymouth Rock chicken and deep-breasted Cornish hen is the most important meat breed. Some varieties are raised for their ornamental appearance, as pets, or for cockfighting.

Chicken pox or **varicella.** Acute infectious disease caused by the herpes zoster virus that also causes Shingles. Usually a disease of childhood, chicken pox is a highly contagious disorder characterized by an itchy rash of blisterlike lesions that appear two to three weeks after infection. When the lesions have crusted over, the disease is no longer communicable. Topical medication is of ten given to relieve the itching.

Chick-pea. Annual plant *(Cicer arietinum)* of the Pulse family, cultivated since antiquity for the edible, pea like seed. Chick peas are used as food for humans and animals. They are boiled or roasted, and have been used as a coffee substitute. Other common names are *ceci*, garbanzo, and gram pea.

Chicory or **succory.** Herb *(Cichorium intybus)* of the Composite family, native to the Mediterranean and widely grown in North America and Europe. The roasted and powdered root is used ass a seasoning and in salads. French endive, or witloof, is the type that is blanched for salad. True endive, or escarole *(C. endivia)*, a salad vegetable since antiquity, is cultivated in broad-and curly-leaved varieties.

Chihuahua. Small Toy Dog; shoulder height, c.5 in. (12.7 cm); weight, 1-6 lb (0.5-2.7 kg). There are two varieties: the smooth, with a short, close-lying glossy coat, and the long coated, with soft-textured, flat, or slightly wavy hair. The coat is usually tan. Probably of Chinese origin, it was introduced into Mexico by Spanish settlers.

Chile saltpetre. *See* sodium nitrate.

Chimpanzee. Black- haired APE (genus pan) of the equatorial forests of centre and W Africa, including the common chimpanzee *(P. troglodytes)* and pygmy chimpanzee *(P. paniscus);* considered the most intelligent of apes. The common chimpanzee is covered with long, black hair and may reach 5 ft. (1.5 m) in height and 150 lb (68 kg) in weight. Captive chimpanzees have been taught to communicate by using Sign Language or a computer console.

Chinchilla. Small burrowing Rodent that lives in colonies up to 15,000 ft (4,270m) high in the Andes of Bolivia, Chile, and Peru. Its soft gray pelt is one of the costliest of all furs, and the wild chinchilla was nearly exterminated before protective laws were passed.

Chip. *See* integrated circuit.

Chitin. Main constituent of the shells of Arthropods. Analogous in structure to Cellulose, chitin contributes strength and protection to the organism. The cell walls of some fungi also contain chitin.

Chiton. Marine invertebrate Mollusk (class Amphineura). The chiton has a low oval body covered dorsally by a slightly convex shell of eight overlapping plates. It clings tightly to hard surface with its broad flat ventral foot; it crawls by means of muscular undulations of the foot. The mouth, located in front of the foot, contains a toothed scraping organ, the radula. Chitons range from 1/2 in. to 12 in. (1.2-30 cm).

Chladni's figures. Regular patterns obtained from fine powder scattered on a vibrating plate. The usual method of obtaining such patterns is to mount a horizontal plate on a central pillar and vibrate it by an electromagnet or mechanical vibrator. The powder collects along nodal lines of the plate where the vibration is least. [After Ernst Chladni (1756-1827), German physicist.]

Chloral (trichlorethanal). A colourless mobile oily liquid, CCL_3 ÇHO, made by the chlorination of acetaldehyde and use in the manufacture of DDT. M.pt.-57.5°C; b.pt. 97.8°C; r.d.1.51.

Chloral hydrate. A colorless aromatic poisonous solid diol, $CCl_3CH(OH)_2$, prepared by the hydrolysis of chloral and used as a sedative and in the manufacture of DDT. M.pt. 57°C; b.pt.96.3°C; r.d.1.91.

Chloramine. A colourless explosive liquid, NH_2Cl, produced by the action of sodium hypochlorite on ammonia. M.pt.-66°C.

Chlorate. Any salt of chloric acid.

Chloric acid. A strong oxidizing acid, $HClO_3$ stable only in aqueous solution. It is formed along with chlorous acid when chlorine dioxide is dissolved in water. When concentrated, it decomposes into chlorine and perchloric acid. Its salts are the chlorates.

Chloride. Any of certain compounds containing chlorine.

Chloride of lime. *See bleaching powder.*

Chlorination. 1. Treatment with chlorine, as in the disinfection of water.

2. A chemical reaction in which one or more chlorine atoms are introduced into an organic compound.

Chlorine (Cl). Gaseous element, discovered in 1774 by Karl Scheele, who thought it was an oxygen compound, and identified as an element by Sir Humphry Davy in 1810. Chlorine is a greenish-yellow, poisonous gas with a disagreeable, suffocating odor. A Halogen, it occurs in natures in numerous and abundant compounds, e.g., sodium chloride (common salt). Chlorine is soluble in water; chlorine water has strong oxidizing properties. Chlorine is used in water purification, and as a disinfectant and antiseptic. Chlorinated hydrocarbons (e.g., DDT) are long-lasting pesticides. Many poison gases contain chlorine. Chlorine compounds used medically include chloroform and chloral hydrate.

Chlorine. Symbol: Cl. A greenish-yellow poisonous nonflammable gas with a pungent odour, occurring in many minerals. Especially in common salt. It is obtained by electrolysis of solutions of sodium chloride. The element is used in the manufacture of paper and many organic chemicals. It is also employed in water treatment (chlorination) and in bleaching textiles. It is a reactive element belonging to the halogen group. A.N. 17; A.W. 35.43; m.pt. -100.98°C; b.pt. -34.6°C; r.d. 2.5 (air = 1): valency 1, 3, 5 or 7.

Chlorine oxide. Any of four oxide of chlorine, all unstable explosive compounds. The lowest is *chlorine monoxide,* Cl_2O, which is a brownish-yellow gas formed by the action of chlorine on mercury II oxide. It is the acid anhydride of hypochlorous acid. M.pt. -116°C; b.pt. 2°C, *chlorine dioxide,* ClO_2 is a reddish-yellow gas made by the action of concentrated sulphuric acid on potassium chlorate at low temperature. It is a mixed anhydride, yielding a mixtures of chlorous and chloric acids in water. M.pt. -59°C; b.pt. 11.0°C. The two higher oxides are both unstable liquids. *Chlorine trioxide,* Cl_2O_6 is made by reaction between chlorine and ozone. M.pt. 3.5°C. *Chlorine heptoxide,* Cl_2O_7, is produced by dehyrating perchloric acid. M.pt.-91.5°C; b.pt. 82° C.

Chlorite. Any salt of chlorous acid.

Chloroacetic Acid. Any of three acids that can be produced by reacting chlorine with acetic acid in the presence of red phosphorous. *Monochloroacetic acid,* $CH_2ClCOOH$, is a colourless crystalline solid. M.pt. 55.6°C (β from); b.pt. 189°C; r.d. 1.37. *Dichloroacetic acid,* $CHCl_2COOH$, is a colourless liquid. M.pt. -4°C; r.d. 1.57. *Trichloroacetic acid,* CCl_3COOH, is a white crystalline solid. M.pt. 57.5°C; b.pt. 197.5°C; r.d. 1.63. All three acids are used as intermediates. The presence of chlorine atoms in the methyl group makes the chloroacetic acids stronger than acetic acid itself, trichloroacetic acid being the strongest.

Chlorobenzene (monochlorobenzene). A colourless mobile flammable liquid, C_6H_5Cl, with an álmond-like odour. It is manufactured by catalytic chlorination of benzene and is used as a raw material in making phenol, aniline, and DDT and as a solvent for paints. M.pt. -45°C; b.pt.132°C; r.d. 1.11.

Chloroethane (ethyl chloride). A flammable gaseous alkyl halide, C_2H_5Cl, with an ethereal odour and a burning taste. It is made by the addition of hydrogen chloride to ethene and used a refrigerant, local anaesthetic, and alkylating agent. M.pt. -137°C; b.pt. 12.3°C; r.d. 0.9.

Chloroethene (chloroethylene, vinyl chloride). An easily liquefied gas, CH_2:-CHCl, with an ethereal odour, manufactured by chlorination of ethene. It is readily polymerized, hence its use in the manufacture of polyvinyl chloride. M.pt. -160°C; b.pt. -14°C; r.d. 0.91.

Chloroform (trichloromethane). A colourless volatile liquid, $CHCl_3$ with a characteristic sweet odour. It is made by reacting chlorinated lime with acetone, acetaldehyde, or ethanol. Chloroform is now principally used in making fluorocarbon compounds. It is also a solvent and anaesthetic. M.pt. -63.5°C; b.pt. 61°C; r.d. 1.48.

Chloromethane (methyl chloride). A colourless flammable gaseous alkyl halide, CH_3Cl, made by the chlorination of methane. Chloromethane is used as a refrigerant and a local anaesthetic. M.pt. -97°C; b.pt. -24°C; r.d., 0.92.

Chlorophyll. Green pigment in plants that gives most their color and enables them to carry on the process of Photosynthesis. Chlorophyll, found in the Choloroplasts of the plant cell, is the only substance in nature able to trap and store the energy of sunlight. The light observed by chlorophyll molecules in mainly in the red and blue-violet parts of the visible spectrum; the green portion is not absorved but reflected, and thus chlorophyll appears green.

Chlorophyta. Division of the plant kingdom consisting of green Algae and the stoneworts.

Chloroplast. Complex, discrete, lens-shaded structure, organelle, contained in the cytoplasm of plant cells. Chloroplasts have submicroscopic, disklike bodies, called grana, that house Chlorophyll and are the central site of the process of Photosynthesis.

Chloroplatinic acid. *See* platinum chloride.

Chloroprene. A colourless liquid, H_2C:-CHCCl:CH_2 made from vinyl acetylene and hydrochloric acid. It is the monomer for the manufacture of neoprene rubbers. B.pt. 59°C; r.d. 0.96.

Chlorosulphonic acid. A colourless or light yellow fuming highly corrosive liquid, $ClSO_2OH$, used in the manufacture of detergents and pharmaceuticals. M.pt. -80°C; b.pt. 158°C; r.d. 1.77.

Chlorous acid. An unstable acid, $HClO_2$ formed together with chloric acid when chlorine; dioxide is dissolved in water. Its salts are the chlorites.

Chocolate. Term for products of the seeds of the Cacao tree, used for making beverages or candy. Chocolate is prepared by a complex process of cleaning, blending, and roasting the beans, which are then ground and mixed with sugar, cocoa butter, and milk solids. A chocolate drink known to the Aztecs came to Europe through Spanish explorers c.1500, and was a fashionable beverage in 17th-and 18th cent. England. In 1765 chocolate was first manufactured in the U.S., now the world's leading producer. The process for making milk chocolate was perfected in switzerland c.1876.

Choke. A coil of wire used as an inductor to impede the flow of an alternating current in a circuit. *Cholera or Asiatic cholera.* Acute infectious diseases caused by the bacterium *Vibrio cholerae.* The bacteria, which are found in fecal-contaminated food and water, produce a Toxin that affects the intestines, causing diarrhoea, severe fluid and electrolyte loss, and, if untreated, death. Treatment consists of administration of glucose and electrolyte solutions. The diseases remains prevalent in regions of Africa and Asia where public sanitation is poor.

Cholesterol. Fatty substance found in vertebrates and in foods from animal sources. A steroid, cholesterol is found in large concentrations in the brain and spinal cord, as well as in the liver, the major site of cholesterol biosynthesis cholesterol is the major precursor of the synthesis of vitamin D and the various steroid Hormones It sometimes crystallizes in the Gall Bladder to form gallstones. Atherosclerosis (see Arteriosclerosis) is associated with deposits of cholesterol inside major blood vessels.

Chord. A straight line joining two points on a curve.

Chordate. Common name for an animal haivng three unique features at some stage of its development: a notochord (dorsal stiffening rod) as the chief internal support, a tubular nerve cord (spinal cord) above the notochord, and Gill slits leading into the pharynx (anterior part of the digestive tract). Grouped in the phylum Chordata, chordates are mainly vertebrates, animals of the subphylum Vertebrata (the Fishes, Amphibians, Reptiles, Birds, and Mammals), in which a backbone of bone or cartilage forms around the notochod; the rest are small aquatic invertebrates of the subphyla Urochordata (Tunicates) and Cephalochordata (Lancelets), in which there are no backbones.

Chorea. Disease causing involuntary jerky, arrhythmic movements of the face, limbs, or entire body. The childhood disease Sydenham's chorea, or St. Vitus' dance, is usually a complication of Rheumatic Fever. The condition develops slowly, sometimes up to six months after the acute infection has occurred, but it resolves completely. In Huntington's chorea, a hereditary disease of adulthood, the jerky movements characteristic of the disorder result from progressive degeneration of the central Nervous system. The disease is inviriably fatal.

Choroid. *See* eye.

Chromate. Any salt of chromic acid.

Christian Science. Religion founded upon principles of divine healing and laws expressed in the acts and sayings of Jesus Christ, as discovered and formulated by Mary Baker Eddy and practiced by the Church of Christ, Scientist. The sect denies the reality of the material world, arguing that sin and illness are illusions to be overcome by the mind; thus, they refust medical help in fighting sickness. Mrs. Eddy's *Science and Health with Key to the Scriptures* is the textbook of the doctrine. The church was founded in 1879 and is centered in Boston.

Chromatic aberration. An aberration of lenses caused by the fact that the refractive index of a material depends on the wavelength of the light, so that different colours are focussed at different points. Images formed in this way are surrounded by fringes of spectral colours. Chromatic aberration can be avoided by using achromatic lenses.

Chromium potassium sulphate (chrome alum). A violet-red efflorescent crystaline alum, $K_2SO_4.Cr(SO_4)_3.24H_2O$, used as a mordant. M.pt. 89°C; r.d. 1.81.

Chromophore. The group responsible for the colour of a dye or other chemical substance. In azo dyes, for example, the chromophore is the azo group, $-N = N-$.

Chromosome. Structural carrier of hereditary characteristics, found in the cell nucleus. The number of chromosomes is characteristic of each species; in sexually reproducing species, chromosomes generally occur in pairs. In Mitosis, or ordinary cell division, each individual chromosome is duplicated and each daughter cell receives all chromosomes, a set exactly like its parent's. In Meiosis, the process by which sex cells (Ovum and Sperm) are formed, each daughter cell receives half the number of chromosomes, one of each pair. A fertilized egg contains two sets of chromosomes, one set from each parent. Chromosomes are made of protein and Nuclec Acid. They represent the linear arrangement of Genes, the units of inheritance.

Chromosphere. The gaseous atmosphere of the sun.

Chromyl chloride. A dark red liquid, CrO_2Cl_2 used as an oxidizing agent in organic synthesis. m.pt. 96.5°C; b.pt. 116°C; r.d. 1.91.

Chronometer. Mechanical instrument for keeping highly accurate time. The perfection of the chronometer in 1759 by the English clockmaker John Harrison allowed navigators at sea to determine longitude accurately for the first time. A marine chronometer is a spring-driven escapement timekeeper, like a watch, but its parts are more massively built and it has devices to compensate for changes in the tension of the spring caused by changes in temperature.

Chronon. The time taken for a photon to travel a distance equal to the diameter of the electron i.e. about 10^{-24} second.

Chrysanthemum. Annual or perennial herb (genus *Chrysanthemum*) of the Composite family, long grown in the Orient. The chrysanthemum is a national flower of Japan and the floral emblem of the Japanese imperial family. The red, white, or yellow flowers range from single daisylike heads to large rounded or shaggy heads. They are commercially important. Innumerable horticultural types exist, most varieties of C. *morifolium*. The pyrethrum, fever few, marguerite, and Daisy belong in the same genus.

Chrysophyta. Division of the plant kingdom consisting of four diverse classes of Algae, the largest and best known of which comprises the Diatoms.

Cicada. Large, noise-producing Insect (order Homoptera) with a stout body, a wide blunt head, protruding eyes, and two pairs of membranous wings. The males have plate like membranes on the thorax, which the vibrate, producing a loud, shrill sound; females are mute. The periodical cicadas (genus *Magicicada*) have the longest life cycles known of any insect, 13 years in one species and 17 in another, but winged adults only live for about one week.

Cinchona or **Chinchona.** Evergreen tree (genus *Cinchona*) of the Madder family,native to mountainous areas of South and Central America and widely cultivated elsewhere for its bark, the source of Quinine and other antimalarial alkaloids. The tree was named for the Countess of Chinchon, said to have been cured of a fever in 1638 by a preparation of the bark.

Cinnabar. Deep-red mercury sulfide mineral (HgS). Used as a pigment, it is principally a source of the metal Murcury. Cinnabar is mined Spain, Italy, and California.

Cinnamon. Tree or shrub (genus *Cinnamomum*) of the Laurel family. Cinnamon spice, obtained by drying the bark of the tropical Ceylon cinnamon (*C. zeylanicum),* has been used since biblical times. *C. camphora* is the source of Camphor.

Cinnabar. A bright red naturally occurring form of mercury II sulphide, HgS, important as the principal ore of mercury.

Cinnamic acid (3-phyenylpropenoic acid). A graphite crystalline carboxylic acid, $C_6H_5CH:CHCOOH$, occurring naturally in certain balsams and essential oils. It exists in two isomeric forms, the *trans* isomer being the common one. Esters of cinnamic acid are used in perfumery and medicine. M.pt. 133°C; b.pt. 300°C; r.d. 1.28 *(trans).*

Circle. The locus of a point that moves so that it is always a fixed distance from a fixed point. The area is πr^2 and the circumference 2 r, where r is the radius. A circle is an ellipse with an eccentricity of zero.

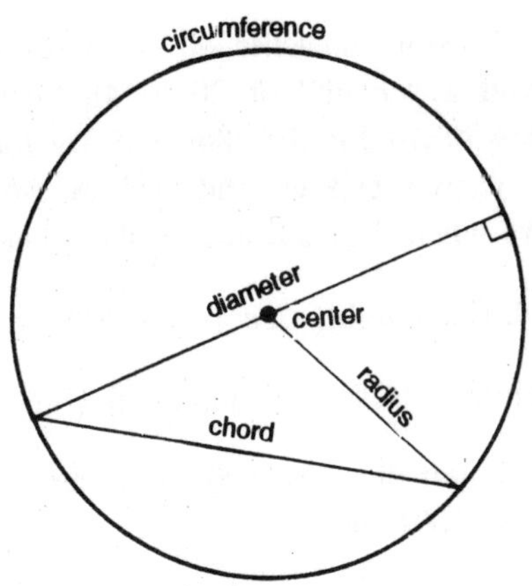

Fig. C-11. Circle.

Circle. Closed plane Curve consisting of all points at a given distance from a fixed point, called the center. A radius of circle is any line segment connecting the center and the curve; the word *radius* is also used for the length r of that line segment. Both the circle itself and its length C are referred to by term *circumference*. A line segment whose two ends lie on the circumference is a *chord;* a chord through the center is a *diameter*. The circumference of circle is given $C = 2\pi r$. The area A bounded by a circle is given by $A = \pi r^2$. In the religion and art of many cultures the circle frequently symbolizes heaven, eternity, or the universe.

Circuit. An arrangement of conductors, resistors, capacitors, etc. through which a current can flow.

Circuit breaker. Electric device that, like a Fuse, interrupts a current in an Electric Circuit when the current becomes too high. Unlike a fuse, which must be replaced after it has been used once, a circuit breaker can be reset after it has been tripped. When a high current passes through the circuit breaker, the heat it generates or the magnetic field it creates causes a trigger to rapidly separate the pair of contacts that normally conduct the current.

Circuit rider. Itinerant preacher of the Methodist denomination who served a "circuit" of 20 to 40 "Appointments". The System was devised in the 18th cent. By John Wesley for his English societies and was adapted in America by Francis ASBURY, where it aided greatly the spread of Methodism.

Circularly polarized light. *See* polarization.

Circular measure. Measure of angles in radians.

Circular mil. A unit of area equal to the area of a circle whose diameter is 0.001 inch.

Circulatory system. A group of organs that transport blood and the substances it carries to and from all parts of the body. In

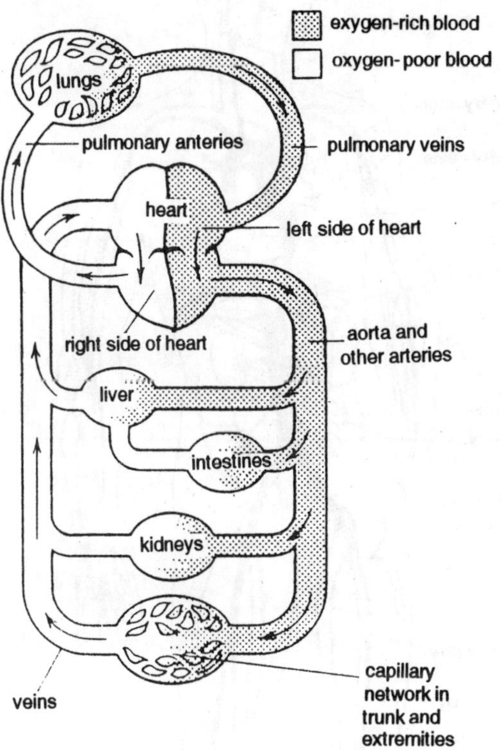

Fig. C-12. *Schematic diagram of the circulatory system.*

humans the circulatory system consists of vessels (Arteries and *Veins)* that carry the blood and a muscular pump, the Heart, that drives the blood. Arteries carry the blood away from the heart; the main arterial vessel, the aorta, branches into smaller arteries, which in turn branch into still smaller vessels that reach all parts of the body. In the smallest blood vessels, the capillaries, which are located in body tissue, the gas and nutrient exchánge occurs—the blood gives up nutrients and oxygen to the cells and accepts carbon dioxide, water, and wastes. Blood leaving the tissue capillaries enters

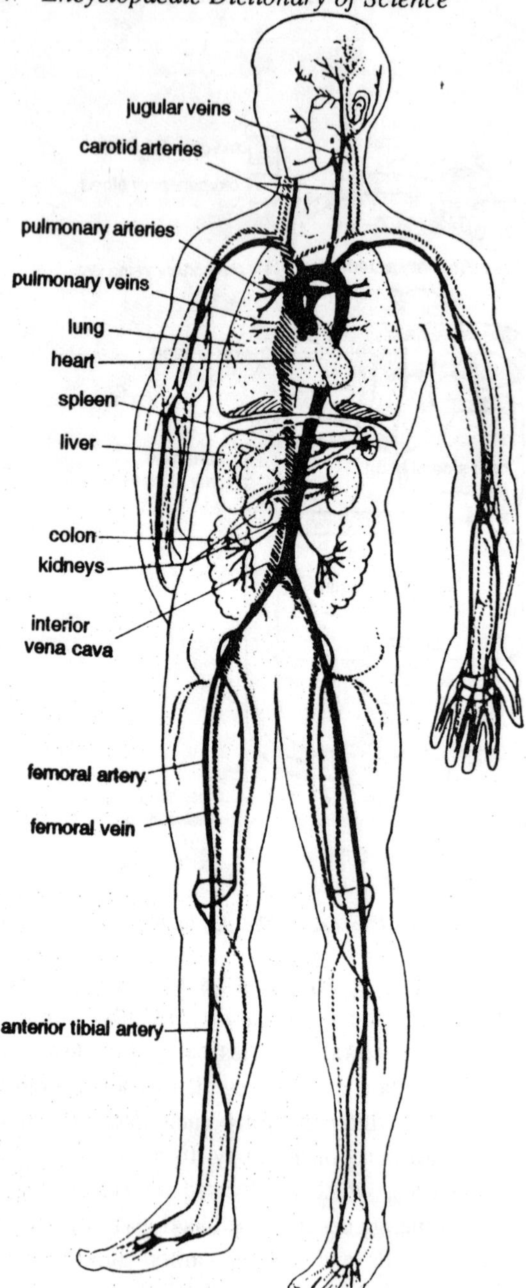

Fig. C-13. *Circulatory system.*

converging large vessels, the veins, to return to the heart. The human heart has four chambers and a dividing wall, or septum, that separates the heart into a right and a left heart. Oxygen-poor, carbon dioxide-rich blood from the veins returns to the right side of the heart. The heart contracts to pump the blood through pulmonary arteries to the Lungs, where the blood receives oxygen and eliminates carbon dioxide. Pulmonary veins return the oxygen-rich blood to the left side of the heart. The left side then pumps the oxygenated blood through the branching aorta and arteries to all parts of the body. The organs most intimately related to the substances carried by the blood are the Kidneys, Spleen, and Liver. An auxiliary system, the Lymphatic System, collects lymph, or tissue fluid, from body tissues and returns it to the blood.

Cirumcentre. The centre of the circle that passes through all three vertices of a triangle. Since this point is equidistant from the vertices. It is the point of intersection of the perpendicular bisectors of the sides of the triangle. *Compare* incentre.

Circumference. The line enclosing any simple closed curve, especially a circle; the length of this line. In the case of a circle of radius r, the circumference is $2\ \pi r$.

Circumpolar. Describing stars or constellations that always remain above the horizon when observed from a given geographical latitude.

Circumpolar star. A star whose apparent daily path on the celestial sphere lies completely above or below an observer's horizon. A star whose declination is greater than 90° minus an observer's latitude will always remain above the horizon for that observer.

Cire perdue [Fr., = lost wax]. Process of hollow metal casting. A plaster or clay model is coated with wax and covered with a mold of perforated plaster or clay. Heat is applied, the wax melts and runs out of the holes, and molten metal (usually bronze) is poured into the space formerly occupied by the wax. When the metal is cool, the mold is broken and the core removed Probably of Egyptian origin, the method was introduced into Greece in the 6th cent. B.C and was used extensively from the 5th cent. The process, employed worldwide, was brought to China c.200 B.C and was later used in casting the Benin bronzes of Africa. The great bronze masterpieces of the Renissance were produced by the cire perdue method.

Cirrhosis, Degeneration of Liver tissue, resulting in fibrosis and nodule formation. Laennec's, or alcoholic, cirrhosis is the most prevalent form, occurring in patients with a history of Alcoholism; other causes include infections, such as Hepatitis, and obstruction of Bile flow. Cirrhosis can result in gastrointestinal disturbances, emaciation, Jaundice, and edema. It is irreversible, but supportive treatment includes diet control, Vitamins, and diuretics (to remove accumulated fluid). Cirrhosis is a common cause of death in the U.S.

Cis. Indicating the position of a group or atom in a molecule adjacent to the position of a specified group or atom: i.e. on the same side of a double bond or central atom. *See* geometrical isomerism.

Cis-trans isomerism. *See* geometrical isomerism.

Citrate. Any salt or ester of citric acid.

Citric acid. a white crystalline solid, $HO_2CCH_2C(OH)(CO_2H)CH_2CO_2H$ with an acid taste, occurring widely in both plant and animal cells. It is extracted by industrial fermentation of molasses or citrus fruits. M.pt. 153°C; r.d. 1.67.

Citric acid cycle or **Kerbs cycle.** Series of chemical reactions occurring in the cells of higher plants, animals and many microorganisms. It is essential for the oxidative metabolism of Glucose and other simple sugars. The reactions take a molecule of citric acid (originating from glucose) through several intermediate products; additional organic molecules are incorporated and citric acid is regenerated. Each cycle releases hydrogen atoms, which are necessary for the next stage of metabolic process to generate chemical energy for the organism.

Citron. Small evergreen tree *(Citrus medica)* of the Orange family, and its fruit, the first of the Citrus Fruits introduced to Europe from the Orient. Citrons are grown in the Mediterranean, West Indians, Florida, and California. The large yellow-green, thin-rinded, and furrowed fruit contains a thick, white, and tender inner rind and a small acid pulp. The juice is used as a beverage and syrup, and the candied, preserved rind is used in confectionery and cookery.

Citrus fruit. Edible fruits of trees of the orange, or Rue, family; almost all native to Southeast Asia and the East Indies. The fruits are rich in vitamin C, citric acid, and sugars; the rind and blossoms yield Essential Oils. Citrus fruits include the Orange, Graperfruit, Lemon, Lime and Kumquat.

Cladding. The process of coating one metal by another, as to prevent corrosion.

Clark cell. A type of primary cell in which the positive electrode is mercury coated with a paste of mercury sulphate, the negative electrode is zinc sulphate solution. It was formerly used as a standard of e.m.f., the value being 1.4345 volts at 15°C. [After William Mansfield Clark (1884-1864), U.S. chemist.]

Clark process. *See* hard water. [After Thomas clark (18601-67), Scottish chemist.]

Classical. Relating to physical theories that do not use either the quantum theory or the theory of relativity.

Classicism. Term meaning clearness, elegance, symmetry, and repose produced by attention to traditional forms; absence of emotionalism, subjectivity, and excess enthusiasm; and, more precisely, admiration of Greek and Roman models. Renaissance writers, for example, looked to Cicero. In England, Francis Bacon in prose and Ben Jonson in poetry strove for classical style. The movement reached its apex with Alexander Pope and the Augustans. In France neoclassicism found its highest expression in the dramas of Pierre Corneille and Jean Raciline. The works of Renaissance painters and of the composers F.J. Haydn and W.A. Mozart particularly reveal the classical impulse. Classics and Romanticism are generally contrary tendencies.

Classification. In biology, the systematic categorization of organisms. One aim of modern classification, or systematics, is to show the evolutionary relationships among organisms. The broadest division of organisms is into kingdoms, traditionally two-Animalia (animals) and Plantae (plants). Fairly widely accepted today are three additional kingdoms; the portista, comprising protozoans and some unicellular algae; the Monera, bacteria; and blue-green algae; and the Fungi From most to least inclusive, kingdoms are divided into the following categories: phylum (usually called *division* in botany), class, order, family, genus, and speckes. The species, the fundamental unit of classification, consists of populations of genetically similar, interbreeding or potentially interbreeding individuals that share the same gene pool (collection of inherited characteristics whose combination is unique to the species).

Clausius, Rudolf Julius Emanuel. (klou'zeoos), 1822-88, German mathematical physicist. He introduced the concept of entropy and restated the second law of Thermodynamics to say that heat cannot of itself pass from a colder to a hotter body. Through investigations of heat, electricity, and molecular physics, he developed the Kinetic-Molecular Theory of Gases and formulated theory of Electrolysis.

Clathrate. A compound formed by physical trapping of molecules of one substance in holes in the crystal lattice of another. Clathrates are produced by crystallizing a mixture of the two substances. Quinol, for example, has holes in its lattice that can trap oxygen methane, methanol, and other compounds with small molecules. The ratio of quinol: compound can vary up to a maximum value of 3:1 Ice can form clathrate compounds with argon, krypton, and xenon. In such compounds no chemical bond is formed between the host compound and the trapped molecule. *See also* zeolite.

Clay. Common name for a number of fine-grained, earthy materials that are plastic when wet. They are easily molded into a form they retain when dry, and become hard and hold their shape when subjected to heat. Chemically, clay minerals are hydrous aluminum silicates with various impurities. Clays are most commonly formed by surface weathering. In the form of Bricks, clay has been indispensable to architecture since prehistoric times. Clays are of great industrial importance, e.g., in the manufacture of tile and pipe. Clay is one of the three principal types of soil; the others are Sand and Loam.

Cleavage. Splitting of a crystal to form smooth surfaces. Cleavage occurs along planes of atoms in the crystal.

Clematis. Herb or vine (genus *Clematis)* of the Buttercup family. The vines are usually profuse and varied bloomers. The Jackman clematis *(C. jackmanii)* is a large purple hybrid; the Japanese clematis *(C. paniculata)* has small white flowers. Other names for clematis are virgin's bower, traveler's joy, leatherflower, and old-man's-beard.

Clilmate. Average weather conditions in an area over long period of time, taking into account temperature, precipitation, humidity, wind, atmospheric pressure, and other phenomena. The major factor governing climate is latitude; this is modified by one or more secondary factors including position relative to land

and water masses, attitude, ocean currents, topography, prevailing winds, and prevalence of cyclonic storms. The earth is divided into climatic zones based on average yearly temperature and average yearly precipitation. The study of climate is called climatology.

Fig. C-14. Clinical thermometer.

Clinical thermometer. A mercury thermometer designed for taking the temperature of the body. It has a constriction in the capillary just above the bulb so that thread of mercury remains in position after reaching its maximum value. It is returned to the bulb by shaking. The instrument is usually calibrated in the range 95-115° F, normal body temperature being 98.4° F.

Clock. Mechanical, electrical, or atomic instrument for measuring and indicating time. Predecessors of the clock were the sundial, the hourglass, and the clepsydra. The operation of a clock depends on a stable mechanical oscillator, such as a swinging Pendulum or a mass connected to a spring, by means of which the energy stored in a raised weight or coiled spring advances a pointer or other indicator at a controlled rate. The heavy and bulky weight-driven clock may first have been built in the 9th cent. The introduction (c.1500) of the coiled spring made possible the construction of smaller, lighter-weight clocks. The Dutch scientist Christian Huygens invented (1656 or 1657) a pendulum clock probably the first. Electric clocks, first made in the late 19th cent., are powered by an electric motor synchronized with the frequency of alternating current. The quartz clock, invented c.1929, uses the vibrations of quartz crystal to drive a synchronous motor at a very precise, rate. An atomic clock (invented 1984), even more precise, is indirectly controlled by atomic or molecular oscillations. Digital clocks and watches dispense with the hour-marked dial and display time as a numerical figure.

Cloisonne. Methods of decorating metal surfaces with enamel. Filaments of metal are attached to the surface of the object, outlining the design and forming compartments that are filled with colored paste enamels. When the piece is heated the enamels fuse with the metal, forming a glossy, colored surface. Probably invented in the Middle East, cloisonne was perfected by the Chinese, Japanese, and French.

Clone. Group of organisms descended from a single individual through asexual Reproduction. Expect for changes in hereditary material due to Mutation, all members of a clone are genetically identical. Experiments resulting in the development of a frog from a cell of an existing animal and the laboratory of fertilization of human eggs have raised questions about the eventual possibility of cloning identical humans from cells of a pre-existing individual.

Cloud. Aggregation of minute of water or ice suspended in the air. Clouds are formed when air containing water vapor is cooled below a critical temperature called the Dew point and the resulting moisture condenses into droplets on microscopic dust particles (condensation nuclei) in the atmosphere. The air is normally cooled by expansion during its upward movement. Clouds may be classified according to appearance, attitude, or composition. Each cloud type is formed by specific atmospheric conditions and is, therefore, indicative of forthcoming weather. *Cirrus* clouds, generally white, and delicate and fibrous in appearance, are the highest clouds and are made of ice crystals. *Stratus* clouds, layered in appearance, are the lowest clouds and are associated with stormy weather. *Cumulus* clouds, which are vertically developed, usually with a horizontal bases and a dome-shaped upper surface, are intermediate in height and are associated with fair weather. Combinations of these clouds names are common. *Cirrostratus clouds,* for example, are high-altitude layered clouds that often indicate rain or snow. Prefixes and suffixes are also used. For example, *nimbus* means rain, and *cumulonimbus* clouds, commonly called thunderheads, indicate rain showers.

Cloud chamber. An apparatus for investigating elementary particles by the tracks they produce on passing through a supersaturated vapour. In the simplest type, the vapour is subjected to a sudden adiabatic expansion, which cools the vapour and produces supersaturation. Particles passing through the chamber leave a stream of irons along their path, which serve as nuclei for the formation of minute drops of liquid, thus producing a visible track. This apparatus is often called the *Wilson cloud chamber* after its inventor.

Clove. Small tropical evergreen tree *(Syzygium aromaticum or Eugenia caryophyllata)* of the Myrtle family and its unopened flower bud, an important spice. The buds, whose folded petals are enclosed in four toothlike lobs of the calyx, are dried and used whole or ground for cooking. Clove oil is used in flavorings, perfumes, and medicines.

Club moss. Living members of Lycopodiopsida, a class of primitive vascular plants that reached their zenith in the Carboniferous period and are now almost extinct. They resemble the more primitive, nonvascular true mosses. Club mosses are usually creeping or epiphytic; many of them in habits moist tropical or subtropical regions. Reproduction is by spores, which are clustered in small cones or borne in the axils of the scalelike leaves. Some species of *Lycopodium,* called ground pine or creeping cedar, resemble miniature hemlocks with flattened fan-shaped branches. Spores of L.*clavatum* are sold as lycopodium powder, or vegetable sulfur, a flammable yellow powder used in pharmaceuticals and fireworks. Species of selaginella are grown as ornamentals, e.g., the resurrection plant.

Clusius column. An apparatus for separating isotopes, consisting isotopes, of a long vertical column many metres high with a heated wire running down its axis. The material to be separated must be gaseous. A radial temperature gradient builds up in the tube between the hot wire and the wall. The lighter molecules tend to diffuse towards the wall and heavier molecules collect around the wire as a result of thermal diffusion. Convection current are set up and the lighter isotope collects at the top of the tube.

Cluster. Any of a number of collections of stars that move together in the same direction. Two main types exist: *open* or *galactic clusters* consist of relatively small numbers of widely spaced young stars and lie chiefly in the spiral arms of the Galaxy. *Globular clusters* are found in the vast spherical halo that encompasses the central nucleus and consist of enormous numbers of closely packed older stars. Open clusters are near neighbours whereas globular clusters are much more remote.

Cluster, star. Group of neighboring stars that resemble each other in certain characteristics that suggest a common origin. Galactic, or open, clusters typically contain from a few dozen to about a thousand loosely scattered stars and exist in regions rich in gas and dust, such as the spiral arms of the galaxy. More than 1,000 galactic clusters, including the Hyades and Pleiades in the constellation Taurus, have been catalogued in the Milky Way. Globular clusters are spherical aggregates of thousands or millions of densely concentrated stars and exist in the outer halo of the galaxy. The brightest of the more than 100 globular cluster so far detected in the galaxy are Omega Centauri and 47 Tucanae, both seem with the unaided eye in the southern skies.

Coagulation. The formation of macroscopic agglomerations in colloids leading, in some cases, to their eventual solidification.

Coal. Fuel substance of plant origin, composed largely of Carbon with varying amounts of mineral matter. Coal belongs to a series of carbonaceous fuels that differ in the relative amounts of moisture, volatile matter, and fixed carbon they contain; the most useful are those containing the largest amounts of carbon and the smallest amounts of moisture and volatile matter. The highest grade of coal is anthracite, or hard coal, which is nearly pure carbon and is used as a domestic fuel. Bituminous coal, or soft coal, with a lower carbon content, is used as an industrial fuel and in making Coke. Lignite and Peat, in descending order, are the lowest in carbon content. Large amounts of coal were formed in the Carboniferous period of geological time (345 to 280 million years ago). It is thought that great quantities of vegetable matter collected and underwent slow decomposition in Swamps similar to present-day peat bogs and in lagoons.The peat that formed was converted to lignite and coal by Metamorphism. The pressure of accumulated layers of overlying sediment and rock forced out much of the volatile matter, leaving beds or seams of compact coal interstratified with shales, clays, or sandstones. Higher grades of coal were produced where the stress was greatest. Major U.S. coal fields are found in Appalachia, the Midwest, the Rocky Mt. region, and along the Gi;f Cpast/ tje cjoef cpa;-producing countries of Europe are germany, Britain, the USSR, Poland, France, and Belgium. Valuable coal deposits also exist in China, India, South Africa, and Australia. *See* also Coal Mining.

Coal gas. A fuel gas produced by the destructive distillation of coal, consisting of mixture of hydrogen (about 50%), methane (35%), carbon monoxide (8%), and other hydrocarbons and smaller amounts of nitrogen, oxygen, and carbon dioxide.

Coal mining. Physical extraction of Coal resources to yield coal; also, the business of exploring for, developing, mining, and transporting coal in any form. Strip Mining is the process in which the earth and rock material overlying the coal is removed to expose a coal seam or bed; the coal is then removed in separate operation. Underground coal mining is the extraction of coal fro below the surface of the earth. The coal is worked through tunnels, passages, and openings connected to the surface form the purpose of removing the coal. Mechanical equipment breaks up the coal to a size suitable for hauling. Alternatively, the coal is drilled, and the resultant holes are loaded with explosives and blasted to break up the coal. To protect miners and equipment, much attention is paid to maintaining and supporting a safe roof system in underground mines.

Coal tar. A black tar produced by the destructive distillation of coal, consisting of a mixture of aromatic hydrocarbons, including benzene, toluene, xylene, and naphthalene together with phenol and some free carbon.

Cobalt. Symbol: Co. A grey hard transition element found in several ores, including smaltite ($(Co,Ni)As_2$) and cobaltite (CoAsS). It is usually obtained by roasting, to give the oxide, followed by reduction with aluminium. Cobalt is a ferromagnetic element and is used in making ferromagnetic alloys as well as in high-speed and stainless steels. It forms two series of compounds: cobalt II *(cobaltous)* and the less stable cobalt III *(cobaltic)* ionic salts as well as many complexes. Cobalt also forms a few compounds in other oxidation states. A.N. 27; A.W. 58.933; m.pt. 1495°C; b.pt. 2870°C; r.d. 8.9; valency 2 or 3.

Cobalt (Co), Metallic element, discovered in 1735 by Georg Brandt. It is silver-white, lustrous, and hard, and can be magnetized. It is combined with other metals in the ores cobaltite and smaltite. Cobalt alloys are used in very hard cutting tools, high-strength permanent magnets, and jet engines. Radioactive cobalt-60 is used in cancer therapy and to detect flaws in metal parts.

Cobalt chloride. A dark red; crystalline solid, $CoCl_2.6H_2O$. The anhydrous salt is blue. Cobalt II chloride (cobaltous chloride) is the only stable chloride of cobalt. M.pt. 86.7°C; r.d. 1.9.

Cobalt oxide. Any of three oxides of cobalt, all used in pigments and ceramics. *Cobalt II oxide* (cobaltous oxide) is green-grey powder, CoO, prepared by heating cobalt II carbonate in nitrogen. M.pt. 1935°C; r.d. 6.5. *Tricobalt tetroxide* (cobalto-cobaltic oxide) is a steel-grey powder, Co_3O_4, obtained by heating other cobalt oxides. It changes to cobalt II oxide when heated above 1000°C, R.d. 6.1, Cobalt III oxide (cobaltic oxide, cobalt black) is a steel-grey or black powder, Co_2O_3, made by heating cobalt III hydroxide. Decomposes above 895°C; r.d. 4.8-5.6.

Cobra. Venomous Snake of the family Elapidae equipped with an inflatable neck hood, found in Africa and Asia. The king cobra *(Ophiophagus hannah)*, or hamadryad, the largest poisonous snake, is found in S.Asia; it may reach a length of 18ft (5.5 m). Others species include the Indian cobra *(Naja naja)* and the Egyptian cobra *(Naja haja)* also called the ASP. The family also includes Coral Snakes.

Coca. Plant (genus *Erythroxylon,* particularly *E. coca)* found mainly in upland reasons and on mountain slopes of South America and in Australia, India, and Africa. Certain South American Indians chew the leaves mixed with lime, which act with saliva to release Cocaine from the leaves. In this form and concentration, the Drug acts as a stimulant. A cocaine-free extract of the leaves is used in some soft drinks. Coca is grown commerically in Sri lanka, java, and Taiwan.

Cocaine. Alkaloid drug derived from Coca leaves, producing euphoria, hallucinations, and temporary increases in physical energy. Prolonged use can cause nervous-system aberration (including delusions), general physical deterioration, weight loss, and addiction. Withdrawal from the drug can produce severe depression.

Cockcroft, Sir John Douglas. 1897-1967, English physicist after serving as fellow professor top natural philosophy at Cambridge, he directed (1946-59) The British Atomic Energy Research Establishment at Harwell. He shared with Ernest Walton the 1951 Noble Prize in physics for their pioneer work in transmuting atomic nuclei by bombarding elements with artificially accelerated atomic particles.

Cocker spaniel. Smallest Sporting Dog; shoulder height, 14-15 in. (35.6-38.1 cm); weight c. 25 lb (11.3 kg). Its silky, flat or wavy coat is moderately long and may be solid color or combination of colors. The breed was developed from English cocker spaniels brought to the U.S. in the 1880s.

Cockle. Heart-shaped marine Bivalve mollusk (order Eulamellibranchia), having ribbed, brittle shells and moving with a jumping motion produced by means of a large, muscluar foot. Most species do not exceed 3 in. (7.5 cm) in length. Cockles burrow in sand or mud in shallow water. Several species are edible.

Cockroach or **roach.** Nocturnal, flat-bodied oval Insect (suborder Blattaria) of the order Orthoptera. Cockroaches have long antennae, long legs adapted for running, and a flat extension of the upper body that covers the head; some species fly. In length they range from 1/4 in. to 3 in.(.6-7.6cm). Cockroaches are not known to be disease carriers. They were extremely abundant during the Carboniferous period (about 350 million years ago) and probably were the first flying animals.

Coconut. Edible fruit of the coco palm tree (*Cocos nucifera*) of the Palm family widely distributed throughout the tropics. The coco palm, which grows to a height of 60 to 100 ft (18-30 m) and has a crown of frondlike leaves, is one of the most useful trees in existence. It is a source of timber and its leaves are used in baskets and for thatch. The coconut itself is a single-seeded nut with a hard, woody shell encased in a thick, fibrous husk. The hollow nut contains coconut milk, a nutritious drink, and its white kernel, a staple food in the tropics, is eaten raw and cooked. Commercially valuable coconut oil is extracted from the dried kernels, called copra, and the residue is used for fodder. Husk fibers are sued in cordage and mats, and nutshells are made into containers.

Cod. Marine bottom-feeding Fish (family Gadidae), among the most important and abundant food fishes. The Atlantic cod *(Gadus morrhua)* averages 10 to 25 lb (4.5 to 11 kg), but specimens up to 200 lb (90 kg) have been reported. The haddock *(Melanogrammus aeglefinus)*, the most important fish in the Atlantic, is smaller, reaching a weight of 30 lb (13.6 kg). Finnan haddie is lightly smoked haddock.

Code. In communications, set of symbols and rules for their manipulation by which the symbols can be made to carry information. For use in a Computer information is encoded as strips of binary digits. For telegraphic work, codes such as the Morse Code are used. All written and spoken languages are codes of some degree of efficiency. Certain arbitrary codes are used to ensure diplomatic and military secrecy of communication. The science of translating messages into cipher or code is termed *cryptography* [Gr. = hidden writing].

Codeine. A white poisonous crystalline alkaloid, $C_{18}H_{21}NO_3$, extracted from opium or made by methylating morphine. Codeine and its derivatives are narcotic and analgesic drugs.

Coefficient. 1. A constant multiplier of a variable in an algebraic expression. In $5x^3 + 2x^2$, 5 and 2 are coefficients.

2. A constant measuring some physical property, as in the coefficient of friction, the coefficient of expansion, the coefficient of viscosity, or the coefficient of restitution.

Coelenterate. Radially symmetrical, predominantly marine invertebrate animal of the phylum Cnidaria (also called Coelenterata), having a three-layered body wall, tentacles, and specialized stinging cells (nematocysts). Members of the phylum have a primitive nervous system and no specialized organs for excretion or respiration.

Coenzyme. Any of a group of relatively small organic molecules that assist Certain Enzymes in their catalytic functions.

Coenzymes participate in chemical reactions catalyzed by enzymes; although often structurally altered in the course of the reaction, the coenzymes are always restored to their original form after it. Important coenzymes include Adenosine Triphosphate, important in the transfer of chemical energy, and Vitamins, vital to a variety of biochemical reactions in the body, including the Krebs cycle.

Coercive force. The magnetic fields strength required to remove the residual magnetism of a ferromagnetic material. If the substance is initially saturated, the coercive force is known as the *coercivity.*

Coffee. Name for an evergreen shrub or tree (genus *Coffea)* of the Madder family its seeds, and the beverage made from them. The mature, red fruit (a drupe) typically contains two seeds, or coffee beans. Varieties of Arabian coffee *(C. arabica)* supply the bulk of the world's supply; Liberian coffee *(C. liberica)* and congo coffee *(C. robusta)* are of some commercial importance. Coffee plants require a hot, moist climate and rich soil. The harvested seeds are cleaned and roasted; heat acts on the essential oils to produce the aroma and flavor. Roasts range from light brown to the very dark Italian roast. Coffee contains Caffeine, a stimulant that can cause irritability, depression, and indigestion if taken in excess. The coffee plant was known before A.D. 1000 in Ethiopia, where its fruit was used for food and wine. A beverage made from ground, roasted coffee beans was used in Arabia by the 15th cent, and by the mid-17th cent. it had reached most of Europe and had been introduced into North America.

Coherent. Relating to electromagnetic radiation that has a single phase. Light from normal sources is not coherent. It consists of a large number of waves all out of phase with one another. Laser radiation, on the other hand, is coherent. Two sources of electromagnetic radiation are said to be coherent if they produce waves that are in phase.

Coherent units. A system of units in which a derived unit is produced simply by multiplying or dividing base units without introducing a numerical factor. The SI units form a coherent system.

Coinage meatals. The groups of metals copper, silver, and gold.

Coke. Hard, grey porous fuel with a high Carbon content. It is the residue left when bituminous COAL is heated in the absence of air. Cock is used in extracting metals from ores in the Blast Furnace.

Cola or **kola.** Tropical tree (genus *cola)* of the sterculia family, native to Africa but grown in other tropical areas. The fruit is a pod containing Caffeine-yielding seeds. Cola nuts are chewed for this stimulant. They are also exported for use in soft drinks and medicine.

Colatitude. *See* polar coordinates.

Colcothar. Red iron oxide, Fe_2O_3.

Cold or **Common Cold.** Viral infection of the mucous membranes of the upper respiratory tract, especially the nose and throat. There are numerous viral organisms that may cause the common cold, although there is no known cure or preventive. Treatment of symptoms includes fluids to prevent dehydration, Analgesics (e.g., Aspirin) to lessen fever, and decongestants to shrink swollen mucous membranes. Some believe that Vitamins in large doses, especially vitamin C, may be helpful in prevention.

Cold cathode. A cathode that emits electrons by field emission rather than by thermionic emission.

Coleus. Tropical plant (genus *Coleus)* of the Mint family native to Asia and Africa. Some, with large, colourful leaves, are cultivated as houseplants.

Colligative properties. Properties of a Solution that depend on the number of solute particles present, but not on the chemical properties of the solute. Colligative properties of a solution include its freezing and boiling points (see States of Matter) and osmotic pressure.

Colligative property. A property that depends on the number of particles of substance, rather than on the nature of the particles. Example are osmotic pressure (see osmosis) and the elevation of boiling points, depression of freezing point, or change in vapour pressure of solvents caused by dissolved substances.

Collimator. A system of slits and (sometimes) lenses for producing a parallel beam of light, other electromagnetic radiation, or particles.

Collodion. A thin film of nitrocellulose obtained by coating a surface with nitrocellulose dissolved in a volatile solvent and allowing the solvent to evaporate.

Colloid. A mixture in which one substance is divided into minute particles (called colloidal particles) and dispersed throughout a second substance. Colloidal particles are larger than molecules but too small to be observed with a microscope; however, their shape and size (usually between 10^{-7} and 10^{-5} cm) can be determined by electron microscopy. In a true Solution the particles of dissolved substance are of molecular size and thus smaller than colloidal particles. In a coarse mixture (e.g., a Suspension) the particles much larger then colloidal particles. Colloids can be classified according to the phase (solid, liquid, or gas) of the dispersed substance and of the medium of dispersion. A gas may be dispersed in a liquid to form a foam (e.g., shaving lather or in a solid to form a solid foam (e.g., Styrofoam). A liquid may be dispersed in a gas to form an Aerosol (e.g., Fog), in another liquid to from an emulsion (e.g., homogenized milk), or in a solid to form a gel (e.g., jellies). A solid may be dispersed in a gas to form a solid aerosol (e.g., dust or smoke in air), in a liquid to form a sol (e.g., Ink), or in a solid to form a solid sol (e.g., certain Alloys), Colloids are distinguished from true solutions by their inability to diffuse through a semipermeable membrane (e.g., cellophane) and by their ability to scatter light (e.g., Tyndall effect).

Colloid. A substance made up of small particles larger than atoms or molecules but too small to be seen by a normal microscope. Usually the particles are dispersed in a medium, giving a *colloidal solution or colloidal suspension.* Colloid particles have diameters in the range 10^{-9} -10^{-5} meter. They can be detected by the ultramicroscope.

Colloid system have a variety of properties depending on the nature of the components. They are broadly classified into *reversible* and *irreversible colloids* depending on whether the components can be separated and the dispersion regenerated by mixing. Another classification is into sols, gels, and emulsions.

Cologarithm. The logarithm of the inverse of a number expressed with positive mantissa. Thus if $\log_{10}(25) = 1.3979$, then $\log_{10}(1/25) = -1.3979$ or $2 + 0.6021$. This is usually written 2.6021 and is the cologarithm of 25. It is used to avoid subtracting mantissae in a calculation.

Color. Visual effect resulting from the eye's ability to distinguish the different wavelengths or frequencies of light. The apparent color of an object depends on the wavelength of the light that it reflects. In white light, an opaque object that reflects all wavelengths appears white and one that absorbs all wavelengths appears black. Any three primary, or spectral, colors can be combined in various proportions to produce any other color sensation. Beams of light are combined "additively", and red, blue, and green are typically chosen as primaries. Pigments, however, combine by a "subtractive" process, i.e., by absorbing wavelengths, and artists generally choose red, blue, and yellow as their primaries. Two colors are called complementary if their light together produces white.

Color blindness. Inability to distinguish colors, an inherited trait occurring almost exclusively in males. Individuals with partially defective color vision (the most common form) have difficulty distinguishing red from green. Those who are completely red-green color-blind see both colors as yellow. Totally color-blind persons see only black, white, and shades of gray.

Color field painting. Abstract art movement begun by painters of the 1960s. Working toward a more intellectual aesthetic than that of Abstract Expressionism, color field painters dealt with their conception of the fundamental formal elements of abstract painting: pure areas of untempered color; flat, two-dimensional space; monumental scale; and the varying shape of the canvas. Painters associated with the movement include Ellsworth Kelly, Morris Louis, Kenneth Noland, and Frank Stella.

Colorimeter. An instrument used in colorimetric analysis.

Colorimetric analysis. A method of quantitative analysis in which the concentration of a coloured solute in a solution is estimated by the intensity of the colour, as by comparing it with standard solutions.

Colour. The visual sensation caused by the wavelength of the light falling on the eye. Colour is detected by the cones in the retina. Colour is also a property of the light itself and of the object reflecting or transmitting the light.

Light of a single wavelength has one of the *spectral colours:* red, orange, yellow, green, blue, indigo, and violet. In general, a colour will be determined by a distribution of wavelengths, but one will predominate: this is its *hue*. Colours with no hue, such as white and black, are *achromatic colours*. *Chromatic colours* can be regarded as a combination of a spectral colour with a certain proportion of white. The *saturation* is the extent to which it is free from white. A third parameter used to describe colours is the brightness or vividness of the colour. In the case of coloured light this is its *luminosity:* in the case of an object or pigment it is called *lightness.*

A difference between coloured lights and coloured pigments also occurs in mixing colours. The former mix directly giving a different colour sensation by an additive process. Pigments mix by a subtractive process. *see also* primary colours, complementary colours.

Colourant. Any substance used to impart colour. Technically. Colourants are classified into dyes, which are applied from solution and pigments, which are insoluble substances.

Colourtron. A type of cathode-ray tube used in colour television, having three election guns, one for each primary colour. The screen is covered with many triangular groups of three coloured phosphor dots, one for each colour.

Coma. Deep state of unconsciousness from which a person cannot be aroused even with the most painful stimuli. It may be caused by severe brain injury, Diabetes, Morphine or Barbiturate poisoning, Shock, or hemorrhage. Treatment is directed at the cause of the condition.

Comb jelly or Sea gooseberry. Solitary marine invertebrate (phylum Ctenophora) having eight radially arranged rows (combs) of ciliated plates (ctenes) on the spherical body surface and specialized adhesive cells (colloblasts) used for capturing Plankton. Comb jellies are carnivorous, bioluminescent, and hermaphroditic. They are weak swimmers but possess a unique sense organ controlling equilibrium (statocyst). Usually transparent, comb jellies vary from 1/4in. (0.6 cm) to more than 1 ft. (30.5 cm) in length and look very much like Jellyfish.

Combination. A selection of a number of entities from a set of entities irrespective of the order of selection. The number of combination of *r* element from a set of *n* is denoted $^{n}C_{r}$ and is given by $n!/r!(n-r)$. See also permutation.

Combustion. The combination of a substance with oxygen to produce heat, and sometimes flame. In a burning gas the heat of reaction is sufficient to maintain the process and light is produced by luminescence of excited molecules or lens or by incandescence of small particles of solid produced in the reaction. In most cases solid and liquids burn when the temperature is high enough for the release of flammable vapours to occur, either by simple vaporisation or by decomposition. Usually combustion involves a complex series of reactions in which carbon is converted to carbon dioxide and hydrogen is converted to water. The term combustion is also used to describe any similar process in which heat and light are generated by chemical reaction, as in the burning of sodium in chlorine. Slow combustion is a process in which a small amount of heat and little or no light is produced by reaction with oxygen.

Comet. Mostly gaseous body of small mass and enormous volume that can be seen from earth for periods ranging from a few days to several months. A comet head, on the average about 80,000 mi (130,000 km) in diameter, contains a small, bright nucleus (theorized to be ice and frozen gases interspersed throughout with particles of heavier substances) surrounded by a coma, or nebulous envelope of luminous gases. As the comet approaches the sun, the particles and gases are driven off, forming a tail as long as 100 million mi (160 million km). Pushed by the Solar Wind, the tail always streams out away from the sun. J.H. Oort hypothesized (1950) a shell of more than 100 billion comets surrounding the solar system and moving very slowly at a distance of as much as 150,000 times the sun-earth distance; a passing star, however, may gravitationally perturb a few closer to the sun. Some, such as Halley's Comet, return periodically.

Comet. Any of a considerable number of celestial objects having a bright head and long conspicuous tail. Revolving about the sun in very ecentric orbits, comets apparently consist of a nucleus composed mostly of ice but also containing dust and other material. When the comet aproaches the sun, the surface of the nucleus begins to vaporize, giving rise to the coma. Under the influence of solar radiation or the stream of ionized gas ejected by the sun, the gas and dust in the coma are pushed out behind (away from the sun) to form the tail, which may be several million kilometres in length.

Common denominator. A number that is multiple of the denominators of two or more fractions.

Communicable diseases. Illnesses caused by microorganisms (bacteria, viruses) and transmitted either directly or indirectly from one infected person or animal to another. Many diseases are spread by airborne microorganisms through contact or proximity (e.g., Influenza, Measles Whopping Cough). Some are spread through contaminated food or water e.g., serum Hepatitis) while others are transmitted by an animal or insect carrier (Malaria, Rabies). Still others are transmitted under special circumstances, such as sexual contact (e.g., Syphilis) or infected instruments or blood transfusion (e.g., serum Hepatitis). Control of communicable disease includes isolation of infected persons, immunization, personal hygiene, and stringent public health and sanitation measures.

Communications satellite. Artificial Satellite that provides a worldwide linkup of Radio and Television transmissions and Telephone service; such a satellite avoids the curvature-of-the -earth limitation formerly placed on communications between ground-based facilities. The first communication satellite was NASA's *Echo 1*, an uninstrumented inflatable sphere that passively reflected radio signals back to earth. Later satellites, starting with NASA's Relay satellites and the American Telephone and Telegraph (AT & T) Company's Telstar satellites, carried with them electronic devices for receiving, amplifying, and rebroadcasting signals to earth. The U.S. Launching (1963) of the first synchronous-orbit satellite (*syncom 1*) paved the way for the formation of the international Telecommunications Satellite Organization, whose successive series of intelsat geostationary satellites have steadily lowered the cost of transoceanic communications. Domestic communications satellite systems, also geostationary, have been developed by Canada, the USSR, Indonesia, and the U.S. Firms Western Union, RCA, and together, AT&T and Comsat. Military systems have been developed by the U.S. and NATO.

Commutative. Denoting operation for which the order of the terms does not affect the result. For example, since x + y = y + x and xy = yx for all real numbers, addition and multiplication are said to be commutative operations.

Commutative law. In mathematics, law holding that for a given operation combining two quantities, the order of the quantities is arbitrary. Addition and multiplication are commutative, thus, a+b = b+a and a x b = b x a for any two numbers a and b. Subtraction and division, however, are not commutative.

Compass. In navigation, an instrument for determining direction. The mariner's compass, probably first used by European seamen in the 12th cent., consists of a magnetic needle freely suspended so that it turns to align itself with the magnetic north and south poles. The *gyrocompass* is a more accurate form of navigational compass, unaffected by magnetic influences, that came into wide use on ships and aircraft during World War II. It consists essentially of a rapidly spinning, electrically driven rotor suspended in such a way that its axis automatically points along the geographical meridian.

Compass. An instrument for indicating the directions north, south, etc. A magnetic compass is simply a magnetized needle pivoted at its centre and free to move in a horizontal plane. It lines up along the earth's lines of force, thus pointing in the direction of magnetic north.

The *gyro-compass* contains a gyroscope rotated by an electric motor and mounted so that it is free to swing in any plane. The axis of rotation is aligned along a north-south direction and maintains this direction irrespective of the movements of the mounting. Unlike the magnetic compass, the instrument is unaffected by local irregularities in the direction of the earth's magnetic field.

Compiler. A computer program for converting information written in a programming language into machine code.

Complement. The angle that, together with a given angle, makes a right angle. Thus 30° is the complement of 60°.

Complementary colours. A pair of colours that give white light when combined together. Orange, for instance, is the complementary colour of blue. Two pigments of complementary colours mix to give black.

Complex. A compound in which atoms or groups are bound to metal atoms or ions by coordinate bonds. An example of a complex is the bright blue copper ammine formed when copper sulphate is dissolved in ammonia solution. It contains the ion $[Cu(NH_3)_4]^{2+}$, in which the four ammonia molecules each donate an electron pair in forming bonds with the Cu^{2+} ion. The species produced is a *complex ion.* Negative complex ions are also formed, as in the ferricyanide ion $[Fe(CN)_6]^{4-}$ and the ferricyanide ion $[Fe(CN)_6]^{3-}$. The ferricyanide ion, for example, can be regarded as produced by coordination of six cyanide ions (CN) to a central iron II ion (Fe^{2+}). The resulting complex ion has an overall charge of 4-. Neutral complexes are also known: the $Pt(NH_3)_2Cl_2$ molecule can be formed by coordination of two (neutral) ammonia molecules and two (negative) chloride ions to a Pt^{2+} ion. Nickel carbonyl and ferrocene are other examples.

The complexes formed by transition metals do not; in general, follow the rule that the metal ion should have the configuration of an inert gas. Often they are paramagnetic because they

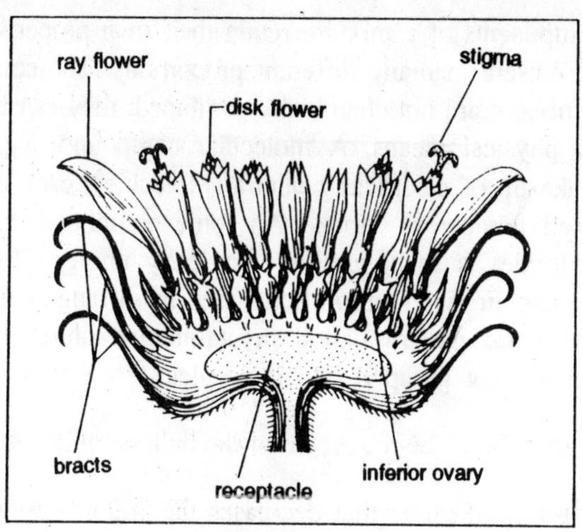

Fig. C-16. A composite flower is composed of many small flowers often of two types (ray and disk). As in the sunflower (above), these flowers are joined by a common receptacle and surrounded by bracts (modified leaves).

Compost. Substance composed mainly of partly decayed organic material, used to fertilize the Soil and increase its Humus content. Usually make from plant materials (e.g., grass clippings, leaves, and Peat), manure, and soil, compost can include chemical Fertilizers and lime.

Compound. A substance composed of two or more elements chemically combined in fixed proportions. Compounds are produced by the formation of chemical bonds between different atoms and they usually have properties that are different from those of their constituent elements. *See* also molecule.

Compound. In chemistry, a substance composed of Atoms of two or more Elements in chemical combination, occurring in fixed, definite proportion and arranged in fixed, definite structures. A compound has unique properties that are distinct from the properties of its elemental constituents and of all other compounds. A compound differs from a mixture in that the components of a mixture retain their own properties and may be present in many different proportions. The components of a mixture are not chemically combined; they can be separated by physical means. A molecular compound, e.g., Water, is make up of electrically neutral Molecules, each containing a fixed number of atoms. An ionic compound, e.g., Sodium Chloride, is make up of electrically charged Ions that are present in fixed proportions and are arranged in a regular, geometric pattern called crystalline structure (see Crystal) but are not grouped into molecules.

Compressibility. The reciprocal of the bulk modulus of a substance.

Compressor. Machine that decreases the volume of air or gas by the application of pressure. Compressor types range from the simple hand Pump and the piston-equipped compressor used in gas stations to inflate tires to machines that use a rotating, bladed element to achieve compression. Compressed air exerts an expansive force that can be used as a source of power to operate pneumatic tools or to control such devices as air Brakes. Air under compression can be stored in closed cylinders.

Compton, Arthur Holly. 1892-1969, American physicist; b. Wooster, Ohio. He was professor of physics at the Univ. of Chicago, where he helped to develop the atomic bomb, and was later professor and chancellor at Washington Univ., St. Louis. For his discovery of the *Compton effect* (the increase in the Wavelengths of x rays and gamma rays when they collide with and are scattered from loosely bound electrons in matter), he shared the 1927 Nobel Prize in physics with Charles Wilson. Compton also make valuable studies of cosmic rays.

Compton effect. The scattering of photons by free electrons, resulting in an increase in the wavelength of the photon (i.e. a decrease in photon energy) and an increase in the velocity of the electron. [After Arthur Holly Compton (1892-1962). U.S. physicist.]

Computer. Any of various devices for processing data according to predetermined instructions. By far the most widely used computers are electronic digital computers which store and process information in digital form. The other main type of computer, the analog computer, accepts a continuous input. Hybrid computers have some of the features of both types.

Computer. A device capable of performing a series of arithmetic or logical operations. The computer is distinguished from a calculating machine, such as an abacus or electronic Calculator, by being able to store a computer program (so that it can repeat its operations and make logical decisions) and to store and retrieve data without human intervention. Computers are classed as analog or digital. An analog computer operates on continuously varying data; a digital computer performs operations on discrete data. An analog computers represents data as physical quantities and operates on the data by manipulating the quantities. In a complex analog computer, continuously varying data are converted into varying electrical quantities and the relationship of the data is determined by establishing an equivalent relationship, or analog, among the electrical quantities. Analog computers are used for a variety of tasks, such as simulating aircraft testing, meteorological tracking and forecasting, and chemical-plant monitoring. Within a digital computer, data are expressed in binary notation (see Numeration), i.e., by a series of "on-off" conditions that represent the digits "1" and "0". A series of eight consecutive binary digits, or bits, is called a byte and allows 256 "on-off" combinations. Each byte can thus represent one of up to 256 Alphanumeric characters. Arithmetic and comparative operations can be performed on data represented in this way and the result stored for later use. Digital computers are used for reservation systems, scientific investigations, data-processing applications, and Electronic Games.

Hardware. The four major physical components, or hardware, of a computer are the central processing unit (CPU), main storage, auxiliary storage, and input/output devices. Computer operations are performed in the CPU, which contains the Logic Circuits for arithmetic and logical operations and for control of the other units that make up a computing system. The CPU also contains the registers, a relatively small number of storage locations that can be accessed faster than main storage and are used to hold the intermediate results of calculations. The main storage is contained in the storage unit, or memory, of the computer. Main storage-once made up of vacuum tubes and later of magnetic cores, each tube or core representing one bit—is now made up of tiny Integrated Circuits, each of which contains thousands of Semiconductors. Each semiconductor represents one bit. Programs and data that are not currently being used in main storage can be saved on auxiliary storage, or external storage. Although punched paper tape once served this purpose, the major materials used today are magnetic tape and magnetic disks. Two areas of continuing research are bubble memories and laser storage. A bubble memory has no moving parts, which improves its reliability; in this technology, a rotating magnetic field moves a magnetic "bubble" (each bubble representing a bit) along loops made of small bars of permalloy. Because the bubbles are so minute in size, huge volumes of data can be stored in tiny spaces. Laser storage involves an array of small holograms, each capable of representing a "page" of binary data. When illuminated by laser light, each page generates a real image of light and dark spots; this falls on a detector array that converts it into electrical signals. Data are entered into the computer and the processed data made available via input/output devices. All auxiliary storage devices are used as input/output devices. For many years, the most popular input/output medium was the punched card. Although this is still used, the most popular input device is now the Computer Terminal and the most popular output device is the high-speed printer. The CPU, main storage, auxiliary storage, and input/output devices collectively make up a system. Multiple systems may be tied together by telecommunications links to form a network.

Software. The computer program, or software, controls the functioning of the hardware and directs its operation. *See* Computer Program.

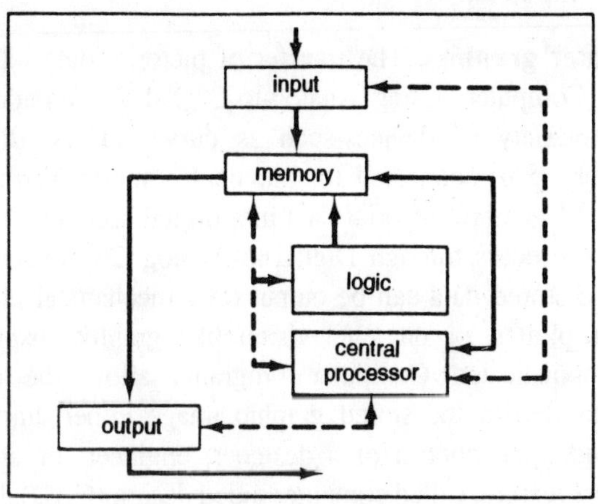

Fig. C-17. Schematic diagram of a computer system: Data flow is indicated by solid lines; control signals are indicated by dashed lines.

History. Although the development of digital computers is rooted in the ABACUS and early mechanical calculating devices, Charles BAGGAGE is credited with the design of the first modern computer. The first fully automatic calculator was the Mark I, or Automatic Sequence Controlled Calculator, begun in 1939 at Harvard by Howard Aiken, while the first all-purpose electronic digital computer, ENIAC (Electronic Numerical Integrator And Calculator), which used thousands of vacuum tubes, was completed in 1946 at the Univ. of Pennsylvania. UNIVAC (Universal Automatic Computer) became (1951) the first computer to handle both numeric and alphabetic data with equal facility. First-generation computers were supplanted by the transistorized computers (*See* Transistor) of the late 1950s and early 1960s,second-generation machines that could perform a million Operations per second. They, in turn, were replaced by the third-generation integrated-circuit machines of the mid-1960s and 1970s. The 1980s have been characterized by the development of the Microsprocessor and the evolution of increasingly smaller but powerful minicomputers, microcomputers, and Personal Computers.

Computer graphics. The transfer of pictorial data into and out of a Computer. Using Analog-to-Digital Conversion techniques, a variety of devices-such as curve tracers, digitizers, and light pens-connected to graphic Computer Terminals can be used to store pictorial data in a digital computer. By reversing the process through Digital-to-Analog Conversion techniques, the stored data can be output on a mechanical plotting board, or plotter, or on a television like graphic display terminal. Sophisticated Computer Programs allow the computer to manipulate the stored graphic image either automatically or under the control of a designer, engineer; or architect. This last feature, called computer-aided design (CAD), has enhanced the productivity of those involved in professions that call for drafting. More recently, as color display terminals and plotters have become readily available, computer graphics have become a popular contemporary art form. *See* also Electronic Game.

Computer programme. A series of actions or instructions that a Computer can interpret and execute; programs are also called software to distinguish them from hardware, the physical equipment used to Data Processing. These programming instructions cause the computer to perform arithmetic operations or comparisons (and then take some additional action bases on the comparison) or to input or output data in desired sequence. Programs are often written as a series of subroutines, which can be used in more than one program or at mote than one point in the same program. Systems programs are those that control the operation of the computer. Chief among these operating system—also called the control program, executive, or supervisor which schedules the execution of other programs, allocates system resources, and controls input and output operations. Processing programs are those whose execution is controlled by the operating system. Language translators decode source programs, written Programming Language, and produce object programs, which are in machine language and can be understood by the computer. These include assebleers, which translate symbolic languages that have a one-to-one relationship with machine language; compiler, which translate an algorithmic-or procedural-language program into a machine-language program to be executed at a later time; and interpreters, which translate source-language statements into object-language statements for immediate execution. Other processing programs are service or utility programs, such as those which "dump" computer memory to external storage for safekeeping and those which enable the programmer to "trace" program execution, and application programs, which perform business and scientific functions such as payroll processing accounts payable and receivable posting, and simulation of environmental conditions.

Computer terminal. A device that enables a Computer to receive or deliver data. Computer terminals vary greatly depending on the format of the data they handle. For example, a simple terminal comprises a typewriter keyboard for input and a typewriter printing element for Alphanumeric output. A similar device includes the keyboard for input and television like screen to display the output. The screen can be a Cathode-Ray Tube or gas plasma panel, the latter involving an ionized gas (sandwiched between glass layers) that glows to form dots which, in turn, connect to form lines. Such display can present a variety of output, ranging from simple alphanumerics to complex graphic images used as design tools by architects and engineers. Under program control, these graphic terminal can rotate the images to show "hidden " parts and even compile parts lists from the design. Other familiar terminals include supermarket check-out systems that deliver detailed printed receipts and use laser scanners to read the vertical stripes of the universal product code printed on food packages, automatic money-dispensing terminals in banks, and sorters that read the magnetic-ink number printed on bank checks.

Concentrated. Denoting a solution that contains a high concentration of solute.

Concentration. The quantity of solute present in a given quantity of solvent.

Concentration. In chemistry, measure of the relative proportions of two or more quantities in a mixture (see Compound). Concentrations may be expressed in a number of ways. The simplest is in terms of a component's percentage by weight or volume. Mixture of solids or liquids are frequently specified by weight-percentage concentrations, whereas mixture of gases are usually specified by volume percentages. Very low concentrations, such as those of various substances in the atmosphere, are expressed in parts per million (ppm). The *molarity* of a Solution is the number of Moles of solute per liter of solution. The *molality* of a solution is the number of moles of solute per 1,000 grams of solvent. The *mole fraction* of a solution is the ratio of moles of solute to the total number of moles in the solution.

Concentric. Denoting circles that have the same centre and lie in the same plane.

Conch. Name for some tropical, marine Gastropods having heavy, spiral shells, the whorls of which overlap each other. The typical gastropod foot is reduced in size and the horny plate (operculum) located at the end of the foot has the appearance and function of a claw. Their shells ranges in color from white to red. Conchs are a valuable food source.

Conchoidal fracture. A type of fracture in which the break has a curved face with a series of concentric rings. Conchoidal fracture occurs in amorphous solids.

Concrete. Structural masonry material made by mixing broken stone or gravel with sand, Cement, and water, and allowing the mixture to harden into a solid mass. Concrete has a great strength under compression, but its tensile strength must be increased by embedding steel rods (reinforced concrete) or Cables under tension (prestressed concrete) within the concrete structural member.

Condensation polymerization. *See* polymerization.

Condenser. 1. A piece of apparatus used for cooling, and thus condensing, a vapour. A common type is the Liebig condenser.

2. *See* capacitor.

Conductance. Symbol: G. **1.** A measure of the ability of a circuit to pass a steady current, equal to the reciprocal of the resistance. It is measured in siemens.

2. A similar quantity used; for alternating-current circuits equal to the real part of the admittance. It is equivalent to $R/(R^2 + X^2)$, where R is the resistance and X is the reactance.

Conductiometric. Denoting an experimental technique that depends on measurements of electrical conductivity. A *conductiometric titration* is one in which the end point is determined by monitoring the conductivity of the mixture as one reactant is added.

Conduction. Transfer of Heat or Electricity through a substance, resulting from a difference in temperature between different parts of the substance or from a difference in electric Potential. Heat may be conducted when the motions of energetic (hottor) molecules are passed on to nearby, less energetic (cooler) molecules, but a more effective method is the migration of energetic free elections. Conduction of electricity consists of the flow of Charges. Metals are thus good conductors of both heat and electricity because they have a high free-electron density.

Conduction. 1. (of heat) Transfer of heat through a material from one point to another point at lower temperature without transfer of mass occurring. In general, conduction occurs as a result of collisions between atoms or molecules-atoms in the hotter region of the material have higher kinetic energies and communicate this energy to neighbouring atoms. Thus the energy is passed on and not carried by movement of atoms as it is in convection. Conduction of heat by this mechanism occurs in most liquids, gases, and nonmetallic solids: these materials are usually poor thermal conductors. In metals, which are usually good thermal conductors, the mechanism is different. The energy is carried by electrons moving through the lattice and colliding with atoms: this is the reason that good conductors of electricity—in both cases the conductivity depends on the motion of electrons.

2. (of electricity.) Transfer of electric charge through a material as a result of an applied electric field. In all cases electrical conduction depends on the presence of charge carriers that can move under the influence of applied electric field. In liquid conductors (electrolytes) the charge is carried by positive and negative ions. In gases the majority of the charge is carried by positive ions and electrons (*see* discharge). In most solids the carriers are electrons although positive holes transport the charge in some semiconductors.

Cone. A surface or solid having a plane base and a vertex above the base, bounded by lines *(generators)* joining the vertex to the base. If the base is a circle the solid is a circular cone has its vertex above the centre of the circle. Such a cone has a lateral surface area Πrh, where h is the slant height, and volume of (1/3)Πr^2s, where S is the attitude.

Configuration. The arrangement of electrons orbiting the nucleus of given atom. Thus the electrons configuration of oxygen is 2,6; these numbers of electrons are present in the first two shells. The configuration can also be written $1s^22s^2p^4$—the 1s, 2s, and 2p denote the subshell and the superscript denotes the number of electrons present.

Confocal. Denoting conic sections that have the same foci.

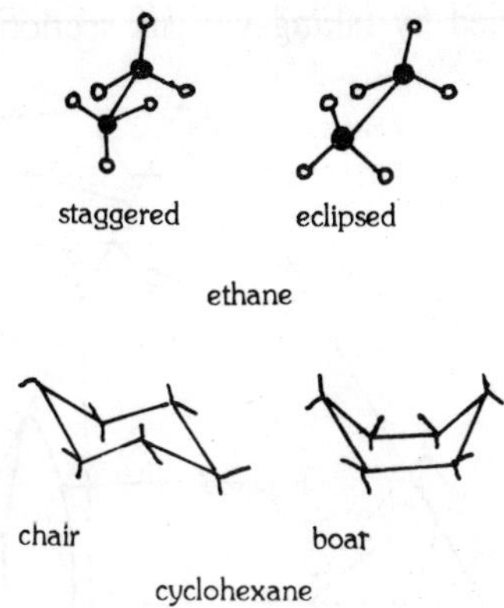

Fig. C-18. Conformations.

Conformation. One of the possible shapes that a molecule may have resulting for rotation of groups about single bonds. The illustration shows two possible conformations for the ethane and cyclohexane molecules.

Conglomerate. A large corporation whose growth comes mainly through acquisition of, or merger with, firms in unrelated fields. Although corporate mergers and acquisitions were common earlier, the modern conglomerate did not emerge until the 1960s, when it quickly became popular among investors. By diversifying, the conglomerate seeks to protect itself against changing markets and economic conditions.

Congruent. Denoting geometric figures that are identical in all respects.

Conic. A type of curve that is the locus of a point that moves so that its distance from a fixed point (the *focus)* is a constant fraction of its distance from a fixed line (the *directrix)* The constant fraction is the *eccentricity, e*, of the conic. Types of conic include the ellipse ($e<1$), the parabola ($e = 1$), and the hyperbola ($e> 1$). The circle is a special case of the ellipse with $e = 0$. Conics are also called *conic section* because they can be obtained by taking various section of a cone.

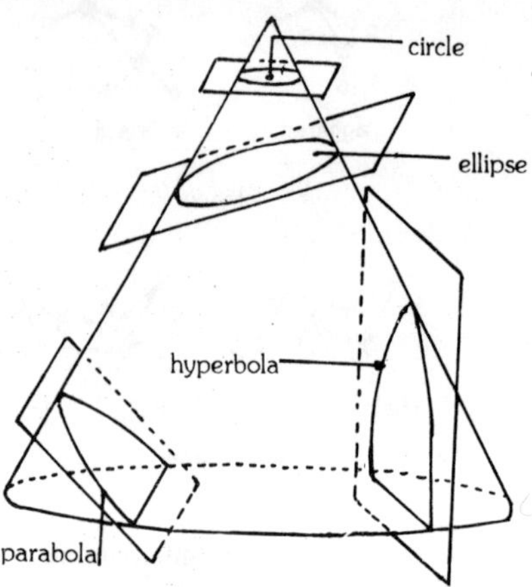

Fig. C-19. Conic sections.

Conic section or **conic.** Curve formed by the intersection of a plane and a right circular cone, or conical surface. The ordinary conic sections are the Circle the ellipse, the parabola, and the hyperbola. When the plane passes through the vertex of the cone, the result is a point, a straight line, or a pair of intersecting straight lines; these are called degenerate conic sections. In Analytic Geometry every conic section is the graph of an equation of the form $ax^2 + bxy + cy^2 + dx + ey + f = 0$, where *a, b, c, d, e,* and *f* are constants and *a,b,* and *c* are not all zero.

Conifer. Tree or shrub of Coniferales, an order of mostly cone-bearing evergreens distributed worldwide. A few, e.g., the Larch are deciduous. Conifers do not have true flowers, but some, e.g., the Yew, have globular fruits. The Pine, Monkey-Puzzle Tree, Cypress, and Sequoia are conifers.

Conjugate. A complex number whose real part of the given complex number and whose imaginary part is equal and opposite in sign to that of the given complex number. Thus a + ib is the conjugate of *a* - *ib*.

Conjugate solutions. Two distinct solutions in equilibrium, each consisting of two liquids, one liquid being the solvent in one of the solutions and the solute in the other. Conjugate solutions are formed under certain conditions from mixtures of partially miscible liquids. Aniline and water, for example, can for two layers. One is a solution of water in aniline and the other a solution of aniline in water.

Conjugation. Interaction occurring in, molecules that contain alternate double and single bonds, characterized by delocalization of electrons in the double bonds. In butadiene, for example, the two double bonds are separated by one single bond and are said to be *conjugated*. Molecules of this type have extra stability in addition to the stability expected in a molecule containing two double bonds and one single bond. This results from the fact that elections are not localized in the double bonds, but can move into the region of the single bond. Benzene is another example of a conjugated molecule.

Conjunction. The coincidence in celestial longitude between two celestial objects. The planets Mercury and Venus are in conjunction twice during their orbits. *Inferior conjunction* occur when the two planets lie between the earth and the sun; *superior conjunction* occurs when the sun lies between the earth and the planets.

Connective tissue. Supportive tissue widely distributed in the body. Consisting mainly of substances certain cells secrete, it contains relatively few cells. The connective tissue that forms Tendons and Ligaments is mainly collagen white, inelastic protein fibers. Cartilage, vital to the skeletal system of vertebrates, is fibrous collagen in a gel; it is firm but flexible. Connective tissue found under the skin and supporting most organs contains collagen and other fibers as well. Bone connective tissue is made up largely of collagen fibers and calcium salt crystals; its structure is strong and rigid. Blood and lymph have a fluid connective tissue matrix.

Conservation laws. In physics, basic laws that maintain that the total value of certain quantities remains unchanged during a physical process. Conserved quantities include Mass (or matter), Energy, linear Momentum, angular momentum, and electric Charge; the theory of Relativity, however, combines the laws of conservation of mass and energy into a single law. Additional conservation laws have meaning only on the subatomic level.

Conservation law. Any principal involving the idea that some property of a system is conserved during changes occurring in the system. Mass-energy, charge, baryon number, lepton number, parity, and similar properties are conserved in certain types of reaction.

Conservation of natural resources. Informed restraint in the human use of the earth's resources. The *conservation,* which came into use in the late 19th cent., referred to the management, mainly for economic reasons, of such valuable natural resources as timber, fish, game, topsoil, pastureland, and minerals. It referred also to the preservation of forests, wildlife, park land, Wilderness, and Watershed areas. Conservation as part of a total approach to use of natural resources was introduced by Pres. Theodore Roosevelt and his chief forester, Gifford Pinchot. They popularized the philosophy, inspired a widespread movement, and gave impetus to much legislation. A Roosevelt-appointed commission published the first inventory of the country's natural resources. Pollution problems and dwindling Energy resources gave rise in the 1960s, and 70s to related movement- Environmentalism - and numerous laws were passed to protect the environment and its resources.. In the early 1980s attempts by the Reagan administration led to open some wilderness areas to resource development led to opposition from both conservationists and environmentalists.

Consistent. Denoting a set of assumptions or propositions that involve no contradiction.

Constant composition, law of. The principlé that given compound always contains the same elements in the same proportions. Water, for example, must always contain hydrogen and oxygen in the ratio 2:16 by weight (i.e. H_2O). The law is one of three laws of chemical combination. It is sometimes called the *law of definite proportions.*

Constellation. In common usage, group of stars (e.g., Ursa Major) that are imagined to form a configuration in the sky; properly speaking, a constellation is a definite region of the sky in which the configuration of stars is contained. The entire celestial sphere is divided into 88 such regions, with boundaries fixed by international agreement along lines of right ascension and declination. The 12 constellations located along or near the ecliptic, or apparent path of the sun through the heavens, are known as the constellations of the Zodiac.

Constellation. A grouping of stars, usually an apparent one caused by the illusion of perspective. The ancients named these shapes after animals and heroes of mythology; Ptolemy listed 48 northern and equatorial constellations in his *Almagest.* With the discovery of the stars of the southern hemisphere, the number has been raised to 88.

Contact lens. Thin lens of plastic (originally glass), worn on the eyeball and used instead of eyeglasses to improve vision or, in certain cases, to correct or treat disorders, The widely used corneal lenses cover just the cornea of the eye and float on the film of tears. Those made of soft plastic can be worn for longer periods of time than the hard-plastic models. Scleral lenses, covering the entire eye, are occasionally used to treat eye injuries or infections.

Contact potential. A small potential difference existing between two different solids in contact. The value is characteristic of the material and varies with temperature.

Contact process. An industrial process for the manufacture of sulphuric acid by the catalytic oxidation of sulphur dioxide. A mixture of sulphur dioxide and air is passed over heated vanadium pentoxide or platinum catalyst. The sulphur trioxide produced is absorbed in concentrated sulphuric acid to give fuming sulphuric acid, which is then diluted.

Continent. Largest unit of land on the earth. The continents ares Eurasia (Europe and Asia), Africa, North America, South America, Australia, and Antarctica. More than two thirds of the continental regions are in the Northern Hemisphere. Continental areas bounded by the sea-level contour comprise about 29% of the earth's surface. All continents contain interior plains, or Plateaus, underlain by the oldest rocks; where exposed to the surface, these rocks are called continental shields, or cratons. The oldest of these rocks dated by radioactivity are 3.8 billion years old,indicating that the cratons formed with the solidification of the earth's crust. The continents have grown by accretion on their edges, where huge plates of crust converge, creating Mountain belts.

Continental drift. The theory that the positions of the earth's continents have changed considerably through geologic time. The first comprehensive theory of continental drift was proposed by the German meteorologist Alfred Wegener in 1912. On the basis of the jigsaw fit of the opposing Atlantic coasts and geologic and paleontologic correlations on both sides of the Atlantic, he advanced the theory that c. 200 million years ago there was one supercontinent, Pangaes, which split into two vast land masses, Laurasia and Gondwanaland. The present continues separated in the next geologic era (the Mesozoic). A plastic layer in the interior of the earth accommodated this process, in which the earth's rotation caused horizontal alterations in the granitic continents floating on the "sea" of the basaltic ocean floors. The frictional drag along the leading edges of the drifting continents created Mountains. Wegener's theory met controversy until 1954, when it was revived by British geophysicists seeking to explain the phenomenon of polar wandering. Since then, the modern theory of Plate Tectonics has evolved from and replaced Wegener's original thesis.

Continuous spectrum. A spectrum that has a continuous distribution of emitted or absorbed wavelengths. Hot solids have continuous emission spectra in the visible or infrared regions. The radiation is produced by changes in vibrational energy of the atoms of the solid.

Continuum. 1. *See* space-time continuum.

2. A continuous region of a spectrum.

Control grid. The grid that is used to control the current in a thennionic valve, distinguished from the screen grid and suppresser grid.

Control rod. *See* fission reactor.

Convection. Transfer of heat by the flow of a liquid or gas. A fluid expands when heated and thus undergoes a decrease in density. The warmer, less dense regions of fluid tend to rise, in accordance with Archimedes' Principle, through the surrounding cooler fluid. If the heat continues to be supplied, the cooler fluid that flows in to replace the rising fluid will also become heated and will rise, setting up a convection current.

Convection. The process in which heat is transferred through a fluid by motion of the fluid. In *natural convention* the motion occurs as a result of a vertical temperature difference. Warm fluid expands and its density decreases, causing it to rise through the surrounding cooler fluid. Its place is taken by cool fluid and cycle *convection currents* are produced.

Convection is the main process of heat transfer in liquids and gases. *Forced convection* is the transfer of heat by an imposed flow of fluid, such as a current of air produced by a fan.

Conventional current. An electric current considered as flowing from a point at high potential to one at lower potential. In fact the charge is usually carried by elections, which flow in the opposite direction.

Convergent. 1. Designating an infinte sequence, the terms of which tend to some limiting value as the sequence progresses. For example, the sequence 1, 1/2, 1/3, 1/4, ... 1/n .... converges to zero as *n* becomes very large.

2. Designating an infinite series that has a finite sum. A series is strictly defined as convergent if the sequence of its partial sums converges. The series 1/2 + 1/4 +1/8 + ... 1/2* + ... is convergent: its partial sums are 1/2, (1/2 + 1/4), (1/2 +1/4 + 1/8), etc., forming a sequence that converges to 1, the sum of the series.

Conversion electron. An electron ejected as a result of internal conversion.

Coolant. A fluid used to transfer heat, as in fission reactor.

Cooper, Peter. 1791-1883, American inventor and industrialist; b. N.Y.C. He built the *Tom Thumb,* one of the earliest locomotives in the U.S., and invented practical devices and processes in the iron industry (e.g., rolling the first iron for fireproof buildings). He invented heavily in the Atlantic Cable and headed North America Telegraph Co., which controlled more than half the telegraph lines in the U.S. interested in providing education for the working classes, he promoted a public school systems in N.Y.C. and founded Copper Union (1859), a free institution of higher learning and a pioneer evening engineering and art school. In 1876 he was the presidential candidate of the Greenback Party.

Coordinate bond (dative bond). A type of covalent bond in which one of the atoms supplies both of the electrons. It can be thought of a combination of transfer and sharing of electrons. Coordinate bonds are also called *semipolar bonds*.

Coordinate geometry. *See* analytic geometry.

Coordinates. Numbers representing the position of a point with respect to reference lines or points. In a plane two numbers are necessary to specify the position: in three-dimensional space three are required. Commonly used coordinate systems are Cartesian coordinates, polar coordinates, and cylindrical coordinates.

Coordination number. *See* complex.

Copernican system. The system of celestial mechanics formulated by the polish monk Nicolaus Copernicus (1473-1543). It superseded Ptolemaic system in respect of the crucial ideas that the earth underwent daily rotation and the sun occupied the central place around which orbited the earth, moon, planets, and stars. The theory was of foundations of modern astronomy.

Copernican system. The first modern European heliocentric theory of planetary motion; it placed the sun motionless at the centre of the solar system with all the planets, including the earth, revolving around it. Copernicus developed his theory (which replaced the Ptolemaic Style) in the early 16th cent. from a study of ancient astronomical records. His system explained Retrograde Motion in a natural way.

Copernicus, Nicholas. 1473-1543, Polish astronomer. After studying law and medicine at the Univ, of Cracow and at Bologna, Padua, and Ferrara, he took up (1512) his duties as canon of a cathedral in Frauenberg, East Prussia. Copernicus laid the foundation for modern astronomy with his heliocentric theory of planetary motion (see Copernican System). This theory, first presented (1512 or earlier) in short form in his unpublished manuscript "Commentariolus," was probably completed by 1530 and was published in his immortal work *De revolutionibus orbium coelestium* (1543).

Copper. Symbol: Cu. A reddish lustrous malleable and ductile transition element found in malachite [$CuCO_3$.Cu-$(OH)]_2$, cuprite (Cu_2O), and chalcopyrite [(Cu,Fe)S_2]. The metal is usually obtained from the sulphide by smelting and electrolytic refining. A certain amount of copper is found native. Copper has a high thermal and electrical conductivity, second only to silver, and is used in electric circuitry and cables. It is also used in alloys such as brass and bronze. The element reacts with oxygen, sulphur, and the halogens and dissolves in concentrated nitric and other oxidizing acids. There are two series of ionic salts: copper I *(cuprous)* and the more important copper II *(cupric)* compounds. The metal also forms many complexes in the I and II oxidation states. A.N. 29; A.W. 63.54; m.pt. 1083°C; b.pt. 2595°C; r.d. 8.96; valency 1 or 2.

Copper (Cu). Metallic element, known to humans since the Bronze AGE. The reddish, malleable, ductile metal is a good conductor of heat and electricity. It has low chemical reactivity. In moist air it forms a patina, a protective greenish surface film. The chief commercial uses are in electrical apparatus and wire, roofing, utensils, coins, metalwork, plumbing, and refrigerator and air-conditioner coils. Copper alloys include Brass and Bronze. Copper compounds are used as insecticides, fungicides, and paint pigments, and in electroplating. Chalcopyrite is the principal ore. Copper is essential for normal metabolism and hemoglobin synthesis in humans.

Copperas. *See* iron sulphate.

Copper carbonate. *See* malachite.

Copper chloride. Either of two chlorides of cooper. *Copper I chloride* (cuprous chloride) is a white crystalline solid. CuCl or Cu_2Cl_2, prepared from copper and copper II chloride solution. It is used as a catalyst, preservative, and fungicide. M.pt. 430°C; b.pt. 1490°C; r.d. 4.14. *Copper II chloride* (cupric chloride) is a brown powder, $CuCl_2$, or green crystalline hydrate, $CuCl_2.2H_2O$, prepared by reaching chlorine with copper. The more stable chloride of copper, it is used as a mordant and a catalyst. M.pt. 620°C; r.d. 3.4.

Copperhead. Poisonous Shake *(Ancistrodon controtrix)* of the E U.S. The copperhead, a Pit Viper, detects its warm-blooded prey by means of a heat-sensitive organ behind the nostril. The head is a pale copper color; the banded, brown body may reach a length of 4 ft (120 cm).

Copper nitrate. A blue crystalline solid, $CuNO_3.3H_2O$, prepared by treating copper or copper II oxide with nitric acid. It is used in dyes and insecticides. Copper II nitrate (cupric nitrate) is the only stable nitrate of copper. M.pt. 114.5°C; r.d. 2.3.

Copper oxide. Either of two basic oxides of copper. *Copper I oxide* (cuprous oxide) is a red-brown powder, Cu_2O, prepared by oxidation of finely divided copper. It is used in ceramics and paints. M.pt. 1235°C; b.p.t. 1800°C; r.d. 5.7. *Copper II oxide* (cupric oxide) is a brown-black powder, CuO, made by heating copper II carbonate. It is used in coloured glazes, insecticides, and as a catalyst. M.pt. 1326°C; r.d. 6.32.

Copper sulphate. A blue crystalline solid *(blue vitriol)*, $CuSO_4.5H_2O$, prepared by addition of dilute sulphuric acid to copper II oxide. It is widely used in pesticides, mordants, and electroplating. The white anhydrous salt is obtained by heating. Copper II sulphate is the only stable sulphate of coppers and is the most important copper salt. R.d. 2.3 (pentahydrate).

Coral. Small, sedentary marine animal of the class Anthozoa, having a horny or calcareous skeleton, also called coral. Most coral form colonies by budding, but solitary corals also exist. In both, the individual animal (polyp) secretes a cup-shaped skeleton around itself; in colonial corals, the skeleton and a thin sheet of living tissue are attached to other individuals. As a colonial coral produces more polyps, the lower members die, and new layers are built up on the old skeletons, forming Coral Reefs. A precious red Mediterranean coral is used for jewelry.

Coral reef. Limestone formation produced by living animals, found in shallow, tropical marine waters. In most, the predominant animals are stony Corals, which secrete skeletons of calcium carbonate (limestone) that build up into a massive formation. Coral reefs are formed only in the tropics where the water is warmer then 72°F (22°C). Coral reefs are classified as fringing reefs (platforms continuous with the shore), barrier reefs (separated from the shore by a wide, deep lagoon), or atolls (surrounding a lagoon). Reefs formed by other organisms are also called coral reefs.

Coral snake. New World poisonous Snake of the same family as the Cobras. The venom of oral snakes, like that of cobras, is especially potent, and the mortality rate among bitten humans is high. Coral snakes are marked by alternating bands of black, red, and yellow.

Core 1. The part of a fusion reactor in which the reaction occurs.

2. The part of a transformer, electric motor, or similar device that constitutes the magnetic circuit. The core is made of iron of other ferromagnetic material and is often laminated to reduce power losses caused by eddy currents.

3. A small semiconductor or a ring of ferrite used to store information in a computer memory.

Coriander. Annual herb *(Coriandrum sativum)* of the Carrot family, cultivated for its fruits. The dried seed is used as a spice, and contains an aromatic oil used as a flavoring, as a medicine, and liqueurs.

Coriolis effect. (For Gaspard Coriolis), tendency for any moving body on or above the earth's surface to drift sideways from its course because of the earth's rotational direction (west to east) and speed, which is greater for a surface point near the equator than toward the poles. In the Northern Hemisphere the drift is to the right of the motion; in the southern Hemisphere, to the left. In most human-operated vehicles, continuous course adjustment mask the Coriolis effect. The Coriolis effect must be considered, however, when plotting ocean currents and wind patterns as well as trajectories of free-moving projectiles through air or water.

Coriolis force. A force considered to act to account for the motion for body as it appears to an observer in a rotating frame of reference. For example, of an object is projected from the centre of a rotating disc along a radius, an observer outside the disc would say that the object was traveling in a straight line above the disc's surface and the disc was rotating underneath. To an observer on the disc, moving with constant angular velocity, the object would appear to be moving in a curved path away from the radius on which he was situated. In fact it would appear to havea tangential acceleration given by $-2(dr/dt)(d\theta/dt)$, where dr/dt is the velocity of the object in a radial direction and $d\theta/dt$ is the angular velocity of the observer. This acceleration is the result of an apparent force.

The Coriolis force, like the centrifugal force, is the "ficticious" effect used to simplify calculations for observation made with respect to rotating systems. It is used to explain certain phenomena caused by the rotation of the earth, such as the motion of air along the isobars of a depression or anticyclone or the motion of water flowing through the plug hole of a bath. [After Gaspard Gustave de Coriolis (1792-1843) French physicist.]

Cork. Protective, waterproof outer covering of the stems and roots of all woody plants, produced by the cork Cambium. Cork cells have regularly arranged walls impregnated with a waxy material, called suberin, that is resistant to water and gas. Cork is buoyant, resilient, light and chemically inert; it is used for bottle stoppers and insulating materials, and in many household and industrial items. The cork Oak *(Quercus suber)*, which has a thick cork layer, is the source of commercial cork.

Corn. In botany, name given to the leading cereal crop of any major region. In England, corn means wheat; in Scotland and Ireland, oats. In America, corn is the grain called maize, or Indian corn *(Zea mays),* a GRASS that was domesticated and cultivated long before Europeans reached the New World. Native Americans raised many varieties (e.g., sweet corn, popcorn, colored corn, and corn for cornmeal). Corn is eaten fresh or ground for meal. It is mainly used as animal fodder. The corn plant consists of the tassel (male flowers) at the top of the plant; the kernels (female flowers) on the cob, enclosed by a leafy husk, beyond which extends the silk (the threadlike stigmas and styles that catch the pollen); and the supporting pithy noded stalk with its prop roots.

Cornea. *See* eye.

Cornflower. Common herb *(Centaurea cyanus)* of the Composite family, a garden flower in the U.S. The longstemmed, blue flower heads have radiating vase-shaped florets that yield a juice used as a dye when mixed with alum. Other names for cornflower include bulebottle, bluebonnet, ragged robin, and bachelor's button.

Corollary. A result following without further detailed proof from a result that has already been proved.

Corona. The gaseous outer region lying above the chromosphere of the sun.

Corona discharge. A luminous electrical discharge produced in air around a conductor at a high critical potential. The discharge, which is often accompanied by a hissing sound, is produced in high electric fields and its formation depends on the diameter and roughness of the conductor. Corona discharges are responsible for power loss in transmission lines. They are also sometimes seen as balls of light around points on church towers, aircraft and ship mats, where the phenomenon is called *St. Elmo's fire* or a *corposant.*

Coronary artery disease. Condition that results when coronary arteries are narrowed or occluded, usually by fibrous and fatty tissue deposits. This condition is the most common underlying cause of Hart Disease and death. Factors contributing to coronary artery disease include Hypertension, high cholesterol levels, and heavy cigarette smoking.

Corposant. *See* corona discharge.

Corpuscular theory. *See* light.

Corrosion. Chemical reaction of a solid, usually a metal, with some substance in the environment leading to eventual destruction or denaturing of the surface. The most commonly observed corroding process is the rusting of iron exposed to air and moisture. Most corrosive reaction are between a solid and a liquid or gas, but solid-solid corrosion also occurs; if two metal so widely different electronegativities are joined together the electrochemical reaction which slowly takes place leads to the destruction of the metallic structure near the junction.

Corrosive sublimate. *See* mercury chloride.

Cort, Henry. 1740-1800, English inventor. He revolutionized the British Iron industry with his use of grooved rollers to finish iron, replacing the process of hammering, and through his invention of the pudding process, which involved stirring the molten Pig Iron reverberatory furnace until the decarburizing action of the air produced a loop of pure metal.

Corticosteroid drug. Any synthetic or naturally derived substance with the general chemical structure of Steroids, used therapeutically to mimic or augment the effects of the hormones produced naturally in the cortex of the Adrenal Gland. Corticosteroids are used to treat Arthritis and other rheumatoid diseases, Asthma Addison's Disease, allergic and inflammatory eye disorders, systemic lupus erythematosus (see Autoimmune Disease), and Hodgkin's and some other forms of cancer. Their anti-inflammatory, itch-suppressing, and vasoconstrictive properties make corticosteroids useful in relieving eczema, psoriasis, and insect bites.

Cortisone. Steroid Hormone whose main physiological effect is on carbohydrate metabolism. It is synthesized from cholesterol in the outer layer, or cortex, of the Adrenal Gland is necessary for the life. Failure of the adrenal gland to synthesize cortisone (Addison's Disease) is fatal unless cortisone is administered. The anti-flammatory effect of the hormone makes it useful in treating asthma and other allergic reactions, arthritis, and various skin diseases.

Corundum. Aluminum oxide mineral (Al_2O_3) occurring in both Gem and common varieties. The transparent gems, chief of which are Ruby and Sapphire, are colorless, pink, red, blue (Oriental Aquamarine) green (Oriental Emerald), yellow, and violet. Common varieties, used as Abrasives (e.g., emery), are blue-gray to brown. Corundum is found in North Carolina, Georgia, Montana, Burma, Sri Lanka, India, Thailand, Republic of South Africa and Tanzania.

Cosmetics. Preparations applied to enhance the beauty of skin, lips, Eyes, hair, and nails. Body paint has been used for ornamental and religious purpose since prehistoric times. Ancient Egyptian tombs have yielded cosmetic jars (kohlpots) and applicators.Thought to have originated in the Orient, cosmetics were used in ancient Greece; in imperial Rome, ladies required the services of slaves adept in applying them. Medieval Europe also used oils and Perfumes brought from the East by returning Crusaders. In the Renaissance white-lead powder and vermilion were used extravagantly. From the 17th cent. Cosmetic recipes abounded, of ten using dangerous compounds, and professional cosmetologists began to appear. After the French Revolution cosmetic virtually disappeared. Since 1900, however, particularly with the growth of the advertising industry, their manufacture and use have grown to huge proportions.

Cosmic background. *See* microwave background.

Cosmic rays. High-energy particles bombarding the earth from outer space. The extraterrestrial origin of cosmic rays was determined (c. 1911) by victor Hess; they were so named by Robert Millikan in 1925. Primary cosmic rays consist mainly of Protons; Alpha Particles; and lesser amounts of nuclei of carbon, nitrogen, oxygen and heavier atoms. These nuclei cause showers of secondary cosmic rays by colliding with other nuclei in the earth's atmosphere. Some cosmic rays have energies a billion times greater then those that can be achieved in particle accelerators; cosmic rays of lower energy, however, predominate. The origin of cosmic rays is still unknown, as are processes by which they are accelerated.

Cosmogony. The branch of cosmology that deals with the origin and evolutionary development of the universe.

Cosmogony. The study of the universe in its entirely, incorporating its origins and development (cosmogony) and a description of it in physical terms as it is observed at present.

Cosmology. Science that aims at a comprehensive theory of the creation, evolution, and present structrue of the entire universe. The Ptolemaic System and the Copernican System are theories that describe the position of the earth in the universe. William Herschel showed that the Milky Way is composed of a vast number of Stars separated by enormous distances. By studying the distribution of star Cluster Harlow Shapley gave the first reliable estimate of the size of our galaxy (c. 100,000 Light-Years) and of the position of the sun within it (c. 30,000 light-years fro the centre). By 1924 astronomers had recognized that some of the Nebulae are not within our galaxy, but are sphere Galaxies at great distances from the Milky Way. After studying the Red Shifts in the spectral lines of the distant galaxies, Edwin Hubble and Milton Humason concluded that the universe is expanding (see Hubble's Law), with the galaxies flying away from one another at great speeds. The big bang theory states that all of the matter and energy in the universe was concentrated in a very small volume that exploded between 10 and 20 billion years ago, and that the resulting expansion continues today. The strongest evidence for the big bang theory is feeble radio background radiation, discovered in the 1960s, that is received from every part of the sky. This radiation has a Blackbody temperature of 3°k above absolute zero and is interpreted as the electromagnetic remnants of the primordial fireball, stretched to long wavelengths by the expansion of the universe. According to another hypothesis, the steady-state theory, the universe expands, but new matter is continuously created at all points in space left by the receding galaxies; this theory now has few adherents.

Containment. The process of confining the plasma to a region away from the walls of a fusion reactor.

Cotangent. *See* trigonometric function.

Cotton. Name for a shrubby plant (genus *Gossypium)* of the Mallow family, for the fibers surrounding the seeds, and for the cloth woven from the spun fibers. Each of the seeds, which are contained in capsules, or bolls, is surrounded by white or cream-colored downy fibers that flatten and twist naturally as they dry. Cotton is tropical in origin but is now cultivated worldwide. It has been spun, woven, and dyed since prehistoric times, most commercial cotton in the U.S. is from *G. hirsutum,* but some is obtained from sea-island and American-Egyptial *G. barbadense* and *G herbaceum.* Cotton is plated annually by seed. diseases and insect pests are numerous, e.g., the boll weevil, responsible for enormous losses, particularly of the highly susceptible, silky-fibered sea-island cotton, which was the leading type of cotton before the advent of this pest . Cotton is separated from its seeds by a Cotton Gin. manufacture of cotton into cloth, Textiles, and yard goods involves carding, combing and Spinning. Cotton is a source of Cellulose products, fertilizer, fuel, plastic reinforcing, automobile tire cord, pressed paper, cardboard, and cottonseed oil (used., e.g., in cooking cosmetics, soaps, candles, detergents, paints, oilcloth, and artificial leather), which is pressed from the seeds. Used in Egypt, China, and India before the Christian era, cotton has long played a significant role in world industry. Britain's need for imported cotton dictated much of its sea-domination policy as an imperial nation, and in the U.S., cotton was a principal economic cause of the Civil War.

Cotton gin. Machine for separating Cotton fibers from the seeds. American inventor Eli Whitney invented (1793) the saw gin, which was especially suited to short-and medium-staple cotton. In a modern gin, which still uses Whitney's basic design, seeds are removed as the fibers are pulled through a grid by a series of circular saws; an air blast or suction carries the clean fibers off the saws to a condenser and, finally, to the baling apparatus.

Coulomb. Symbol: C. the derived SI unit of electric charge, equal to the charge transported by a current of one ampere in one second. [After Charles Augustin Coulomb (1736-1806), french physicist.]

Coulomb, Charles Augustin de. 1736-1806, French physicist. he is known for his work on electricity, magnetism, and friction, and he invented a magnetoscope, a magnetometer, and a torsion balance that he employed in determining torsional elasticity and in establishing Coulomb's Law. The unit of electric charge, the coulomb, is named for him.

Coulombmeter. *See* voltameter.

Coulomb's law. The principal that the force between two point charge is proportional to the magnitude of the charges and inversely proportional to the square of the distance between them. Thus, two charges q_1 and q_2 placed a distance *d* apart each experience a force Cq_1q_2/d^2. The constant C depends on the system of units used. In SI units the force is in newtons, the distance in meters, and the charge in coulombs and the force in free space is given by $q_1q_2/4\pi\varepsilon_0 d^2$, where ε_0 is the electric constant. Sometimes a similar equation is used for the force between charges in a medium: $F = q_1q_2/4\pi\varepsilon_r\varepsilon_0 d^2$, where ε_r is the relative permittivity. However, this extension of Coulomb's law is not strictly correct. The law was used as the basis of the definition of the electrostatic unit of charge (esu). In these units, $F = q_1q_2/Kd^2$, where F is expressed in dynes, d in centimeters, and q_1 and q_2 in esu. The constant K, is the permittivity and it has the value unity for free space.

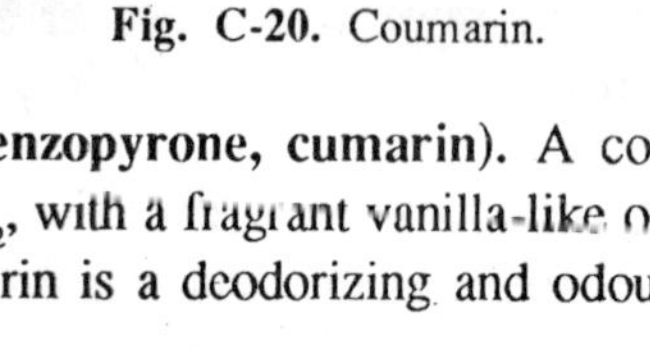

Fig. C-20. Coumarin.

Coumarin (1,2-benzopyrone, cumarin). A colourless crystalline solid, $C_9H_6O_2$, with a fragrant vanilla-like odour and a burning taste. Coumarin is a deodorizing and odour-enhancing agent.

Counter. Any of various instruments for detecting and determining number of particles and photons. A counter consists of a detector which produces an electric pulse as a result of the impact of and individual particle. Associated with this in an electronic circuit for counting the pulses formed. Counters are extensively used in nuclear physics, cosmic ray studies, etc. see Geiger counter, Cerenkov radiation, crystal counter, scintillation counter, ionization chamber.

Counter-glow. *See* gegenschein.

Couple. Two parallel forces acting together in opposite direction at different points, thus producing a turning effect. See moment, torque.

Coupling. Interaction between different effects in system.

Covalent bond. A type of chemical bond in which atoms are held together by shared pairs of electrons. Hydrogen atoms, for example, have a single election and the carbon atom has an electron configuration of 2, 4. One carbon atom can for four covalent bonds to four hydrogen atoms by donating one electron to each bond, the other electron being supplied by the hydrogen atom: the electrons in each bond, have opposite spins. As a result of this process eight electrons are moving in the field of the carbon nucleus, giving it the stable electron configuration of the inert gas neon. Similarly, each hydrogen atom has two electrons, giving it the electron configuration of helium.

Covalent compound of this type are composed of individual molecules. In the solid state they form molecular crystals in which the molecules are held together by fairly weak forces. For this reason they are low-melting and volatile and tend to be insoluble in water, dissolving instead in nonpolar solvents. Each shared pair of electrons constitutes a *single bond.* It is also possible for two atoms to bond by sharing two *(double bond)* or three *(triple bond)* pairs of electrons. Theories of covalent bonding are based on the use of molecular orbitals.

Cover crop. Green temporary crop grown to prevent or reduce Erosion and to build up the soil's organic content. Grains, Legumes (e.g., Clover), and some vegetables are used. Cover crops often are the first means used to rehabilitate neglected or misused land.

Cowbird. New World Bird of the Blackbird and Oriole family. Most lay eggs in the nests of smaller birds, especially vireos,

Sparrows, and flycatchers the host bird usually incubates the eggs and feeds the intruder. Cowbirds feed chiefly on insects, following cattle to capture the insects they stir up in grazing.

Coyote or **prairie wolf.** Small, swift Wolf *(Canis latrans)* found in deserts, prairies, open woodlands, and brush country. Resembling a medium-sized dog with a pointed face, thick fur, and a black-tipped, bushy tail, the coyote is common in the central and W U.S. and ranges from Alaska to Central America and the Great Lakes; it is occasionally seen in New England. Considered dangerous to livestock, coyotes are killed each year by the thousands.

Crab. Chiefly marine animal (a Crustacean) of the order Decapoda, with an enlarged cephalothorax covered by a broad, flat shell (carapace). Extending from the cephalothorax are five pairs of legs, the first pair bearing claws (pincers). Crabs have a pair of eyes on short, movable stalks and many mouthparts. They tend to move sideways, but can move in all directions. The blue crab of the Atlantic coast of the U.S. is a source of food, marketed after molting as the soft-shell crab. The giant Japanese spider crab, with a 1-ft wide (30-cm) carapace and legs about 4 ft (122 cm) long, is the largest arthropod.

Crab Nebula. Diffuse gaseous Nebula surrounding an optical Pulsar in the constellation Taurus. It is the remnant of a Supernova explosion in 1054 and is a strong emitter of radio waves and X ray

Crab nebula. The expanding gaseous nebula M1 in the constellation Taurus. It is identified as the remnant of a supernova observed by Chinese astronomers in 1054. Growing at a rate of 1100 km s^1, it lies at a distance of some 1000 light years from the sun. The diffuse central parts of the nebula give rise to a continuous spectrum, while a bright line emission spectrum is produced by the luminous filaments surrounding the nucleus. The nebula contains a pulsar.

Cranberry. Low, creeping, evergreen bog plant (genus *Oxycoccus)* of the Heath family. The tart red berries are used for sauces, jellies, pies, and beverages. The native American, or large, cranberry (*O.* or *V macrocarpus)* is commercially cultivated. The unrelated high-bush cranberry, or cranberry tree, is in the Honeysuckle family. Cranberries are often classified in the blueberry genus *Vaccinium.*

Crane. Large, wading marsh Bird of the Northern Hemisphere and Africa, related to the Rails. Cranes are known for their loud, trumpeting call and rhythmic mating dances. The North American whooping crane, a white bird almost 5 ft (152 cm) tall, is nearly extinct; the gray sandhill crane, which winters W of the Mississippi, is also becoming rare.

Crayfish or **crawfish.** Edible freshwater Crustacean smaller than, but structurally similar to, the Lobster. Crayfish are found in ponds and streams in most parts of the world except Africa. They are scavengers and grow to 3 to 4 in. (7.6-10.2 cm) in length. Crayfish are usually brownish green; some cave-dwelling forms are colourless.

Cremationl. Disposal of a corpse by fire, widely practiced in some societies, notably India and ancient Greece and Rome. The rise of Christianity ended the custom of Europe, but it revived as cities grew and cemeteries became crowded. Belief in the purifying properties of fire, a desire to light the way of the dead, and fear of the return of the dead may have been among early motivations for cremation. The practice is not sanctioned by the Roman Catholic Church or traditional Judaism.

Creep. Slow permanent deformation of a metal caused by continuous stress.

Creosote. Either of two mixtures distilled from tars. The creosote used for preserving wood is *coal-tar creosote,* a yellow or brown mixture of a aromatic hydrocarbons and some phenols (including cresols and naphthols). *Woodtar creosote* is used in pharmacy for the treatment of bronchitis. It is obtained by distilling wood tar, especially from beechwood, and is largely a mixture of phenols.

Cresol. Any of three isomers, CH_3-C_6H_4OH, obtained from coal tar. A mixture of cresols is used as a germicide.

Cricket. Slender, chirping, hopping Insect of the family Gryllidae. Most have long antennae, muscular hindlegs, and two paris of fully developed wings. The males produce the characteristic song by rubbing specialized structures of their front wings together; both sexes have auditory organs on the forelegs. Common field crickets of the U.S. are brown to black and about 1 in. (2.5 cm) long.

Crinoid. Member of the class (Crinoidea) of marine invertebrate animals (Echinoderms) that includes the sea lilies and feather stars. Most sea lilies remains stalked and sessile for their entire lives, while feather stars break off the stalk and become mobile as adults. All crinoids live with their oral side upward with a ring of arms encircling the mouth. Most have 10 arms, but some sea lilies have up to 40 and some feather stars up to 200.

Critical angle. The smallest angle of incidence at which electromagnetic radiation suffers total internal reflection.

Critical damping. *See* damping.

Critical mass. The minimum mass of fissile material for which a chainreaction is self-sustaining.

CRO. *See* cathode-ray oscilloscope.

Croaker. Carnivorous, elongate, spiny-finned Fish of the family Sciaenidae, including the drum and weakfish (sea trout), so called because of their croaking or grunting noises. Croakers are found in the sandy shallows of temperate and warm seas.

The drums are the largest and noisiest croakers- up to 150 lb (68 kg) in the common drum—and include the commercially important red drum, or channel bass. The weakfish, named for their easily torn flesh, include the common weakfish, or squeteague, or the Atlantic coast and the spotted weakfish.

Crochet. Construction of fabric by interlocking loops of thread or yarn using a hook. The chain stitch is used to cast on a foundation chain of the desired length, into which successive rows of stitches are worked. All stitches and patterns derive from single and double crochet. Like Knitting, crochet produces a fabric that stretches. When a fine hook and thread are used, crochet can be used to produce a form of Lace.

Crocodile. Carnivorous Reptile (order Crocodilia) found in tropical and subtropical regions, distinguished from the Alligator by greater aggressiveness, a narrower snout, and a long, lower fourth tooth, which protrudes when the mouth is closed. Crocodiles have flattened bodies, short legs, and powerful jaws. The saltwater crocodile is often 14 ft (4.3 m) long, while the Nile, American, and Orinoco crocodiles are commonly 12 ft (3.7 m) long.

Crocus. Perennial herb (genus *Crocus)* of the Iris family, natives to the Mediterranean and SW Asia. Crocuses usually bear a single yellow, purple, or white flower and have small, grasslike leaves. One species, Saferon, is cultivated commercially as a yellow dyc. The unrelated meadow saffron, or autumn crocus, is in the Lily family, and the wild crocus, or pasqueflower, is in the Buttercup family.

Cro-Magnon man. Biologically modern human being (species *Homo sapiens),* existing 40,000-35,000 years ago (see Man, Prehistoric). Skeletal remains were first found (1868) in France and then in other parts of Europe. Like Neanderthal Man, who he superseded, Cro-Magnon man stood erect (6 ft/ 180 cm); he had a high forehead, developed chin, and large brain. His advanced upper Paleolithic Culture (see Stone

Age) produced flint and bone tools, shell and ivory jewelry, and elegant polychrome cave paintings of great vitality.

Crompton, Samuel. 1753-1827, English inventor (1779) of a Spinning machine that for the first time allowed the production of fine strong cotton yarns. Crompton's mule spinner, or muslin wheel, combined the best features of Richard Arkwright's water frame and James Hargeaves's spinning jenny.

Crookes radiometer. An instrument for detecting infrared radiation. It consists of an evacuated bulb containing four vertical vanes of thin metal mounted on a freely-rotating horizontal wire cross piece. Each vane is blackened on one side and polished on the other. The black faces absorb more radiation than the polished faces and, at low pressure, molecules rebounding from these faces have more momentum than those from the bright faces. The arrangement is such that the vanes rotate in the presence of a source of heat radiation.

Crookes, Sir William. 1832-1919, English chemist and physicist. Noted for his work on radioactivity, he invented the spinthariscope (used to make visible the flashes produced by bombarding a screen with the alpha rays of a particle of radium), the radiometer (used to measure the intensity of radiant energy), and the Crookes tube (a highly evacuated tube through which is passed an electrical discharge). He founded (1859) *Chemical News* and discovered the element Thallium.

Cross-linkage. A chain of atoms linking two longer chains in a polymer.

Cross product. *See* vector product.

Cross section. Symbol: σ. A measure of the probability of some specified process occurring when a beam of particles or photons interacts with nuclei, atoms, or molecules, expressed as an effective area. For example, if gas-phase atoms are ionized by a beam of electrons then only a certain proportion of the total number of collisions leads to ionization. The ionization cross section is the area that each atom would present to the beam if every collisions led to ionization. The cross section is not the physical size of the atom: it depends on the process and on the energy of particles. It is often used in particle physics to describe the interaction of particles with atomic nuclei.

Crotonaldehyde. A colourless flammable liquid, $CH_3CH:CHCHO$, used as an accelerator in producing rubbers and as a lachrymatory war gas. M.pt.-69° C; b.pt. 102°C; r.d. 0.85

Crotonic acid (but-2-enoic acid). A white crystalline carboxylic acid, $CH_3CH:CHCOOH$, made by the oxidation of crotonaldehyde and used in synthesizing resins, polymers, plasticizers, and drugs.

Croup. Acute obstructive inflammation of the larynx in young children, usually ages three to six. Symptoms include difficulty in breathing and a high-pitched, barking cough due to swelling or spasm. The cause may be an infection, Allergy, or obstruction by swallowed object. Treatment is directed at the cause.

Crow. Black Bird (family Corvidae) related to the Raven, Magpie, and Jay and among the most intelligent of birds. Known for its throaty "caw", the American, or common, crow, about 19 in. (49 cm) long with a wingspan of over 3 ft (92 cm), destroys harmful insects and rodents.

Crown glass. *See* optical crown.

CRT. *See* cathode-ray tube.

Crucible. A dish-shaped vessel suitable for heating chemicals to high temperatures

Crustacean. Invertebrate animal (class Crustacea), an Arthropod.

Primarily aquatic, they have bilaterally symmetrical segmented bodies covered by a chitinous exoskeleton that is periodically shed. In most, the head and thorax are fused as a cephalothorax and protected by a shield-like carapace. The head typically has two pairs of antennae, three pairs of biting mouthparts, and usually one medial and two lateral eyes. Thoracic appendages are often modified into claws and pincers with gills at their bases. Crustaceans include Shrimp, Crayfish, Lobsters, Crabs, and Barnacles.

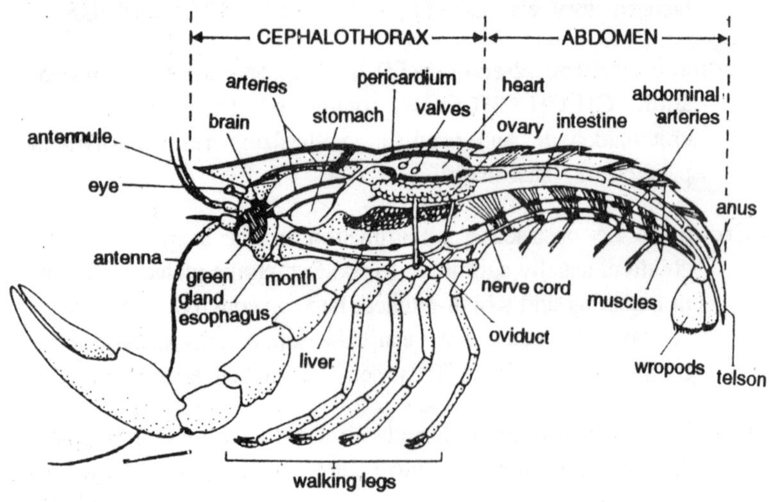

Fig. C-21. Crustacean: Internal anatomy of a female crayfish, a representative crustacean.

Cryogenics. The branch of physics concerned with the production of very low temperatures and the study of phenomena occurring at these temperatures.

Cryohydrate. A crystalline solid containing water and a salt in definite proportions, obtained by freezing solutions of certain salts. Cryohydrate are eutectic mixtures.

Cryolite. A colourless or white naturally occurring form of sodium aluminium fluoride, Na_3AlF_6, used in the production of aluminium.

Cryolite or **kryolite.** Sodium and aluminum fluoride mineral (Na_3AlF_6), usually pure white or colorless but sometimes tinted pink, brown, or even black and having a waxy luster. Cryolite is used principally as a flux in the smelting of Aluminium, but it is also a source of soda, aluminum salt fluorides, and hydrofluoric acid. Discovered (1794) in Greenland, it occurs almost nowhere else.

Cryometer. A thermometer suitable for use at low temperatures.

Cryoscope. Any instrument or apparatus for determining a freezing point.

Cryoscopy. The determination of freezing points, especially in the determination of molecular weights by freezing point depression.

Cryostat. A vessel or enclosure that can be maintained at a constant low temperature.

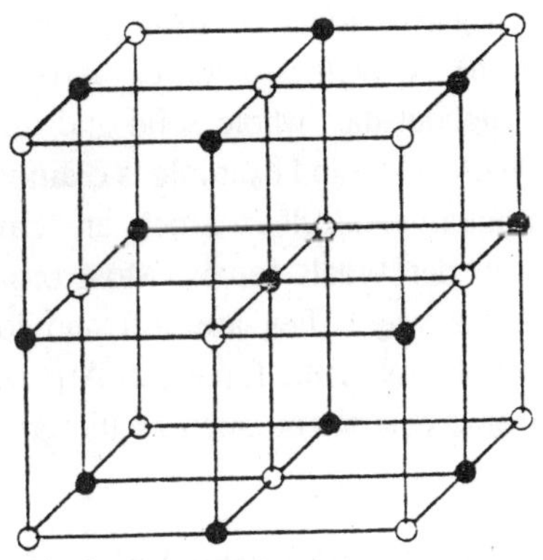

Fig. C-22. Sodium chloride.

Cryptography. Science of translating messages into ciphers or codes. The science of breaking codes and ciphers without the key is called cryptanalysis. Cryptology is the science embracing both cryptography and cryptanalysis. The beginnings of cryptography can be traced to the Hieroglyphics of early Egyptian civilization (c.1900 B.C.) ciphering ; has always been considered vital for diplomatic and military secrecy. The widespread use of computers and data transmission in commerce and finance in making cryptography very important in these fields as well. Recent successes in applying certain aspects of computer science to cryptography have stimulated progress in this area, and seem to be leading to more versatile and more secure systems in which the ciphering or coding is implemented with sophisticated digital electronics.

Crystal. A piece of a solid substance with a regular geometrical shape (the crystal habit) caused by a regular arrangement of the atoms, ions, or molecules making up the solid. The particular type of arrangement is the *crystal structure* and any solid that has such a regular array is said to be *crystalline*, otherwise it is amorphous. If the regular pattern is continued with the same orientation throughout the solid, the piece of material is a *single crystal:* many crystalline substances are polycrystalline. Ionic crystals are composed of a three-dimensional array of positive and negative ions: an example is sodium chloride (see illustration). *Covalent crystals* have covalent bonds extending throughout the whole solid and may be thought of as very large molecules: and example is diamond. In *molecular crystals* individual molecules (which are covalent) are held together by van der Waals forces. Most organic compounds crystallize in this way. They are soft and low-melting as a result of the weakness of the forces. **2.** A piezoelectric crystal used in electronic equipment, such as the crystal pick-up and crystal oscillator.

Crystal. Solid body bounded by natural plane faces that are the external expression of a regular internal arrangement of constituent atoms, molecules, or ions. The particles in a crystal occupy position with definite geometrical relationships to each other, forming a kind of scaffolding called a crystalline lattice. On the basis of its chemistry and the arrangement of its atoms, a crystal falls into one of 32 classes; these in turn are grouped into seven systems according to the relationships of their axes. Differences in the physical properties of crystals sometimes determine the use to which they can be put in industry.

Crystal counter. A type of counter in which a crystal of certain materials is connected across a high potential difference. Each particle of ionizing radiation hitting the crystal makes it momentarily conducting and produces a pulse of current.

Crystal form. An arrangement of crystal planes in three dimensions, having a particular symmetry. Real crystals are characterized by having either one particular form or a combination of forms. The form of a crystal is distinct from its crystal habit.

Crystal habit. The external appearance of a crystal. The habit depends on the internal pattern of atoms and also on the way in which the crystal has grown.

Crystalline. Having a crystal structure rather than an amorphous structure.

Crystallite. One of the small grains in a polycrystalline solid.

Crystallography. The study of crystals and crystal structures. See X-ray crystallography.

Crystalloid. A chemical compound that is crystalline and soluble in water. The term is now obsolete, being formerly used to distinguish compounds that undergo dialysis from *colloids* (which do not dialyse).

Crystal microphone. A type of microphone in which vibrations of a diaphragm are used to vibrate a piezoelectric crystal, thus producing a varying e.m.f.

Crystal oscillator. A type of oscillator circuit in which a fixed frequency is produced by the vibrations of a quartz crystal. The frequency of such a circuit can be constant to one part in 2×10^7. Crystal oscillators are used as standards of frequency and in the quartz clock.

Crystal pick-up. A pick-up used on record players in which vibrations produced by the record groove are converted into a varying e.m.f. by a piezoelectric crystal.

Crystal plane. A plane of atoms, ions, or molecules in the lattice of a crystal.

Crystal spectrometer. A type of spectrometer for use with X-rays, in which the radiation is dispersed by diffraction from the face of a single crystal. See X-ray diffraction.

Crystal system. Any of seven classes into which all crystals are classified, depending on the lengths and angles between the edges of the unit cell. The table shows the characteristics of the seven crystal systems. Sometimes the rhombohedral system is included under the hexagonal system.

Cube. **1.** A regular solid with six equal square faces.

2. The third power of a number: $x^3 = x \times x \times x$.

Cubic. *See* crystal system.

Cuckoo. Bird of the family Cuculidae, widely distributed in temperate and tropical regions. Cuckoos are slender-bodied, long-tailed birds with down-curved bills, pointed wings, and dull plumage. Typical American cuckoos are the black-billed an yellow-billed cuckoos *(Coccyzus americanus)*, known for their low, chuckling call notes. The roadrunner *(Geococcyx californianus)*, a good cuckoo of the southwest deserts, speeds over the ground at up to 15 mi (24.1 km) per hr.

Cucumber. Herbaceous vine *(Cucumis sativus)* of the Gourd family. Its greenish, generally cylindrical fruit is eaten fresh and picked.

Cumarin. *See* coumarin.

Cumene (isopropyl benzene). A colourless flammable liquid hydrocarbon, $C_6H_5CH(CH_3)_2$, distilled from petroleum and used as a fuel additive and solvent. M.pt. -96°C; b.pt. 153°C; r.d. 0.86.

Cumin or **cummin.** Low annual herb *(Cuminum cyminum)* of the Carrot family, long cultivated in the Old World for the aromatic, seedlike fruits. Cumin is an ingredient of curry powder. It yields an oil used in liqueurs and veterinary medicines. Related to the Caraway, cumin has similar uses in cooking.

Cupellation. A method of refining precious metals such as gold and silver by melting them in a shallow porous vessel (*a cupel*). Impurities such as lead are oxidized and absorbed into the walls of the cupel.

Cupric. *See* copper.

Cuprite. A bright red lustrous naturally occuring oxide of copper, Cu_2O, used as an ore.

Cuprous. *See* copper.

Curie. Family of French scientists. *Pierre Curie.* 1859-1906, scientist, and his wife, *Marie Sklodowska Curie,* 1867-1934, chemist and physicist, b. Poland, married 1895, are known for their work on radioactivity and radium. Pierre discovered (1883) and, with his brother Jacques, investigated piezoelectricity (a form of electric polarity) in crystals. Following Antoine Becquerel's discovery of Radioactivity, Marie began to study Uranium, a radioactive element found in pitchblende. Together, the Curies discovered Polonium and Radium and determined their atomic weights and properties. For their work on radioactivity, they shared with Becquerel the 1903 Nobel Prize in physics. Marie Curie became the first person to be awarded a second Nobel Prize when she received the 1911 chemistry prize for the discovery of polonium and radium. The French scientists *Frederic Joliot Curie.* 1900-1958, formally Frederic Joliot, and *Irene Joliot-Curie.* 1897-1956, daughter of pierre and Marie Curie, were married in 1926. They received the 1935 Nobel prize in chemistry for artificially producing radioactive substances by bombarding elements with alpha particles. The Joliot-Curies investigated (1940) the chain reaction in nuclear fission. In 1946 they helped to organize the French atomic energy commission, of which Frederic was the first chairman (1946-50).

Curie. Symbol: Ci. A unit of radioactive disintegration rate equal, for a specified radioisotope, to the amount of the isotope that produces $3.7 x 10^{10}$ disintegrations per second. After Marie Curie (1867-1934), French physicist born in Poland.]

Curie's law. The principle that the susceptibility of a paramagnetic substance is inversely proportional to its thermodynamic temperature (T): $x = C / T$, where C is a constant, the *Curie constant*, for the material. A modification of this law, the *Curie-Weiss law*, has the from $x = C /(T - \theta)$, where θ is a fixed temperature. Ferromagnetic materials lose their ferromagnetism above a fixed temperature (the *Curie point*) and display paramagnetic behaviour. They then follow the Curie-Weiss law, θ being the temperature at the Curie point.

Curium. Symbol: Cm. A silvery transuranic actinide element made by neutron bombardment of americium. It was originally known before americium, being made by bombardment of plutonium with alpha particles. The most stable isotope, ^{248}Cm, has a half-life of $4.7 X 10^5$ years.

Curium. (Cm), Synthetic element, first produced by alpha-particle bombardment of plutonium-239 by Glenn Seaborg and colleagues in 1944. It is a silvery metal and a very radioactive element in the Actinide Series. Curium accumulates in bones and disrupts red blood cells. see Element (table); Periodic Table.

Curlew. Large shore Bird (genus *Numenius)* of both hemisphere, usually brown and buff in color with a downcurved bill. The long-billed curlew (*N. americanus)* is found in the West.

Currant. Northern shrub (genus *Ribes)* of the Saxifrage family. The gooseberry bush belongs to the same genus. The tart black, white or red currant berries and the purple gooseberries are both eaten fresh or used in preserves, sauces, and pies. Dried currents were used by native Americans in making pemmican, a level food. Today's commercial "dried currant" is a raisin. Because gooseberries and currant are a host to blister Rust, their cultivation is discouraged.

Current. Symbol: I. A flow of electric charge. The magnitude of the current is equal to the charge carried per unit time and is measured in amperes. In metals the charge is transported by electron: in semiconductors the carriers may be positive holes. By convention the current in a circuit is taken to flow from regions of potential potential to regions of negative potential. This is known as a *conventional current*. In fact, the electrons, which carry the charge, flow in the opposite direction.

Current balance (ampere balance). An instrument for determining an electric current by directly measuring the force produced between conductors. Several types exist: in the commonest a flat horizontal coil is suspended from one arm of a balance so that it lies between two fixed flat horizontal coils. The current to be measured is passed through the three coils in opposite directions and the value can be calculated by the force produced, as determined by the balance. This type of instrument is used in defining the value of the newton.

Current density. Symbol: *j*. The electric current flowing per unit area at a point in conductor, beam of charged particles, etc.

Cursor. A transparent slider with a vertical hair-line on it, used on slide rules to facilitate reading of the scales.

Curvature of field. An aberration of optical systems in which the images of off-axis points lie on a curved surface rather than on a plane.

Curve. In mathematics, the path of point moving in space. In Analytic Geometry a plane curve, i.e., a curve that lies in one plane, is usually considered as the graph of an equation or function (for some examples, see Conic Section). A skew, twisted, or space curve is one that does not lie all in one plane, e.g., the helix, a curve having the shape of a wire spring.

Cuttlefish. Mollusk (order Sepiodea) that has 10 tentacles, 8 bearing muscular suction cups on their inner surface and 2 longer ones for grasping prey. A Cephalopod, the cuttlefish has a reduced internal shell embedded in the mantle; in some there is a degenerate internal shell of lime called cuttlebone. The body is short, broad, and flattened with lateral fins, similar to the Squid. When disturbed, cuttlefish eject a dark ink, which hides them from predators.

Cuvier, Georges Leopold Chrestein Frederic Dagobert, Baron. (kuvya), 1769-1832, French naturalist. A pioneer in comparative anatomy, he originated a system of zoological classification based on structural differences of the skeleton and organs. His reconstruction of the soft parts of fossils deduced from their skeletal remains greatly advanced paleontology, and he identified and named the flying reptile pterodactyl. He rejected evolutionary theory in favor of Catastrophism.

Cyanamide. 1. (carbodiimide). A colourless deliquescent crystalline solid, HN:C:NH, made by reaction between carbon dioxide and hot sodamide. M.pt. 41°C.

2. *See* calcium cyanamide.

Cyanate. *See* cyanic acid.

Cyanic acid. A volatile liquid, HN:C:O. the compound with this formula is sometimes called *isocyanic acid* and the term *cyanic acid* is used for the isomer HOC:N. This is also known as *fulminic acid.* Covalent derivatives of both acids are known. *Isocyanates* have the formula RNCO and *cyanates* have the formula ROCN. Similarly ionic compounds can be formed from both acids and again there is some confusion in naming the salts. Compounds containing -NCO ion are called *cyanates or isocyanates.* Those containing the $^{-}$OCN ion are *cyanates or fulminates.*

Cyanide. Salt or ester of hydrogen cyanide (HCN) formed by replacing the hydrogen with a metal or a radical. The most common and widely used (those of sodium and potassium) are employed as insecticides, in making pigments, in metallurgy, and in gold and silver refining. Most cyanides are deadly poisons that cause respiratory failure. Symptoms include a breath odor of bitter almonds, dizziness, convulsions, collapse, and, often, froth on the mouth.

Cyanide process. A method of extracting gold from its ores by dissolving it in potassium cyanide.

Cyanogen A colourless highly poisonous flammable gas, $(CN)_2$, with an odour of almonds . It can be prepared by heating mercury cyanide and has been used as a fumigant, war gas, and rocket fuel. M.pt.-28° C; b.pt.—21°C; r.d.1.81.

Cybernetics. Term coined by Norbert Winner to refer to the general analysis of control systems and communication systems in living organisms and machines. Analogies are drawn between the functioning of the brain and nervous system and that of the Computer and other electronic systems. Cybernetics overlaps the fields of Automation, computing machinery, Information Theory, and neurophysiology.

Cycad. Palmlike plant of the Cycadales, an order of mostly topical and subtropical cone-bearing evergreens. Cycads, known from the Permian period , are the most primitive of the living Seed-bearing plants. Some have tuberous, underground stems and crowns of leathery, glossy, fernlike leaves arising from ground level; others have high, columnar stems. Some Cycads, e.g., the fern palm of the Old World tropics and the Australian nut palm, bear edible, nutlike fruits. Florida Arrowroot, or sago, is a starch from the pith of the coontie, or sago palm *(Zamia floridens)* The order comprises nine genera with fewer than a hundred species.

Cyclamate. Organic salt used as an artificial sweetener. Its use in foods was banned in 1969 when it was reported that massive doses caused cancer in rats. There is no evidence that cyclamate is associated with cancer in humans.

Cycle. A sequence of changes occurring in a system and eventually restoring the system to its original state. The term is often used to describe one such set of changes in a system that is periodically varying. In an alternating current or voltage or in a wave, the cycle is the sequence of values that is periodically repeated. In one cycle an alternating current may increase to a maximum, decrease to zero, increase to a maximum in the opposite direction, and decrease to zero again. The time taken for a cycle is its period.

Cycle per second. *See* hertz.

Cyclic. 1. Denoting a chemical compound that has a ring of atoms in its molecular structure. Cycle compounds can be aromatic as with benzene, or alicyclic, as with cyclohexane. When used to describe a class of compounds, that the term is usually taken to mean that the functional group forms part of the ring. For instance, a *cycle either* is a compound such as dioxan, in which the -O-group is bound to other àtoms in the ring.

2. Denoting a plane geometric figure that can be incribed in a circle.

Cyclization. The conversion of an organic compound containing a chain into a compound containing a ring. Many method exist, most of them involving standard types of reaction in which one end of the chain reacts with the other and *ring closure* occurs.

Cyclohexane. A colourless mobile flammable liquid, C_6H_{12}, prepared by the distillation of petroleum or the hydrogenation of benzene. Cyclohexane is used in manufacturing nylon and as a solvent. M.pt. 6.5°C; b.pt. 81°C; r.d. 0.78.

Cyclohexanol. A colourless oily cyclic alcohol, $C_6H_{11}OH$, with a camphor-like odour. It is manufactured by the hydrogenation of phenol and used in making nylon and celluloid and as a solvent. M.pt. 23°C; b.pt. 161°C; r.d. 0.94.

Cyclohexanone. A colourless or pale yellow oily cyclic ketone, $C_6H_{10}O$, made by the oxidation of cyclohexanol. Cyclohexanone is used in making adipic acid and caprolactam and is widely used solvent. M.pt. -47°C; b.pt. 157°C; r.d. 0.95.

Cycloid. A curve that is the locus of a point on the circumference of a circle as the circle rolls along a straight line. It is a special case of a trochoid.

Cyclone. Region, often called a "low", of central atmospheric pressure relative to the surrounding pressure. The resulting pressure gradient, combined with the Coriolis Effect, cause air to circulate about the centre, or core, in a counterclockwise direction north of the equator and a clockwise direction south of it. The frictional drag on near surface air moving over land or water causes it to spiral inward toward lower pressures; this movement is compensated for near the center by rising currents, which are cooled by expansion when they reach the lower pressures of higher altitudes. The cooling, in turn, characteristically increases the relative Humidity greatly and produces cloudiness. An *anticyclone* has the opposite characteristics: a "high", or region of high central pressure relative to the surrounding pressure; clockwise circulation north of the equator and counterclockwise circulation south of it; descending and diverging air that is warmed by compression as it encounters higher pressure at lower altitudes; and characteristic low humidity and little cloudiness. Both cyclones and anticyclones move across the land at speeds of 500 to 1,000 mi/day(800 to 1,600 km/day).

Cyclopentadiene. A colourless liquid hydrocarbon, C_5H_6, obtained by cracking petroleum. With electropositive metals it forms the cyclopentadienyl anion, $C_5H_5^-$.

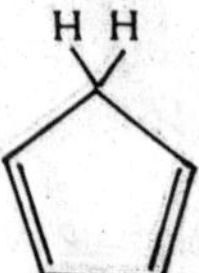

Fig. C-23. Cyclopentadiene.

Cyclotron. A type of particle accelerator in which the particles move in spiral paths under the influence of a uniform vertical magnetic field and are accelerated by an electric field of fixed frequency. The magnetic field is produced by a powerful electromagnet. The particle move inside two hollow D shaped metal electrodes (the *dees)* separated by a small gap.

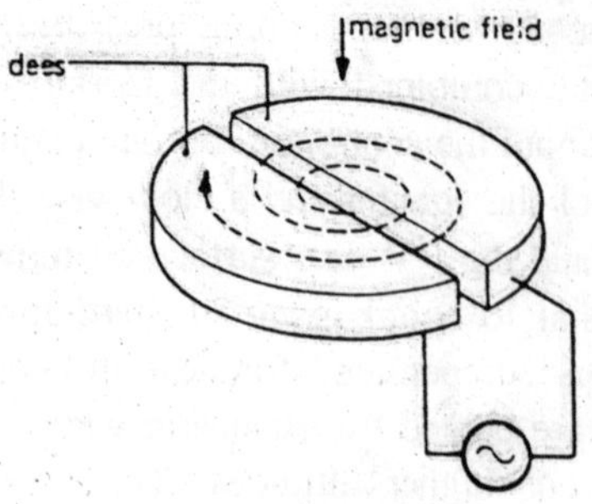

Fig. C-24. Cyclotron.

The time taken to describe a semicircle is Π m/Be, m being the mass and e the charge of the particle, and is independent of the path length, thus allowing the particles to be accelerated by an electric field of fixed frequency applied across the gap between the dees. The frequency is such that the particles are accelerated every time they cross the gap. When they reach the edge of the device they are deflected away onto the target. The energy obtained in the cyclotron is limited by relativistic effects. At high particle energies the increase in mass cause the time taken to complete a semicircle to be significantly larger., The motion of the particles then falls out of step with the frequency of the electric supply and a limiting energy is reached. Typical energies obtained are of the order of 25 MeV: higher energies are obtained in the synchrocyclotron and the synchrotron.

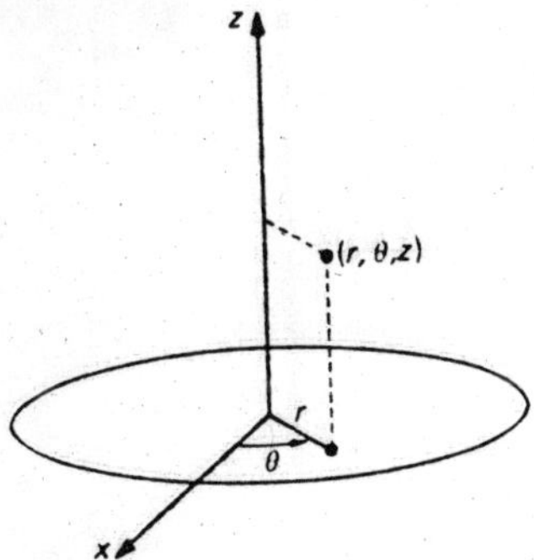

Fig. C-25. Cylindrical coordinates.

Cylindrical coordinates. Coordinates used to locate the position of a point in space by its polar coordinates in one plane and its perpendicular distance from that reference plane.

Cystine. A white crystalline amino acid, $HOOCCH(NH_2)CH_2SSCH_2SSCH_2CH(NH_2)$-COOH.

Cystic fibrosis. Inherited disorder of exocine Glands, affecting infants and children. It is characterized by secretion of a thick, sticky mucus from the exocine glands, which blocks the ducts of these glands (especially pancreas, lungs, and liver). Symptoms include a distended abdomen, diarrhoea, malnutriction, and repeated incidence of respiratory infections. Treatment consists of a low-fat, high protein diet, Vitamins, pancreatin, and Antibiotics to ward of infection.

Cytochrome. Class of heme-caontaining protein, discovered in 1886. Cytochromes play a vital role in Respiration, transporting electrons (as hydride ions) generated by oxidative biochemical reactions, and in the Citric Acid Cycle. An oxygen molecule is the ultimate electron acceptor. Various oxidized Coenzymes are involved in the electron transport process.

Cystaine. A white crystalline amino acid, $HOOCCH(NH_2)CH_2SSCH_2CH(NH_2)\text{-}COOH$.